Notes on

Quantum Mechanics

Notes on

Quantum Mechanics

A Course Given by ENRICO FERMI
at the University of Chicago

Second Edition, with Problems Compiled
by ROBERT A. SCHLUTER

THE UNIVERSITY OF CHICAGO PRESS
Chicago & London

Enrico Fermi was born in Rome in 1901. In 1927, he became Professor of Theoretical Physics at Rome University. In 1938, he was awarded a Nobel Prize for his identification of new radioactive elements and for his discovery of nuclear reactions effected by slow neutrons. Also in 1938, because of the political state of Italy, he and his family moved to New York, where Fermi formed a research group at Columbia University. In 1941, he and his group moved to the University of Chicago and created, at the end of the next year, the first controlled, self-sustaining nuclear reaction. Soon after, Fermi went to Los Alamos, where the Manhattan atomic bomb project was established. After World War II, he returned to Chicago to resume teaching and his research into radioactivity until his untimely death in 1954.

Robert A. Schluter, a doctoral student of Enrico Fermi's (1954), is professor emeritus of physics at Northwestern University.

The University of Chicago Press, Chicago 60637
The University of Chicago Press, Ltd., London

First edition published 1961. Second edition published 1995.
Printed in the United States of America

01 00 99 98 97 96 95 6 5 4 3 2 1

ISBN: 0-226-24381-8 (paper)

Library of Congress Cataloging-in-Publication Data

Fermi, Enrico, 1901–1954.
Notes on quantum mechanics : a course given by Enrico Fermi at the University of Chicago / with problems compiled by Robert A. Schluter, -- 2nd ed.
p. cm.
1. Quantum theory. I. Schluter, Robert A. II. Title.
QC174.F44 1995
530.1'2--dc20 95-3383
CIP

This book is printed on acid-free paper.

CONTENTS

PREFACE TO THE FIRST EDITION

Enrico Fermi taught courses on quantum mechanics on many occasions.

In the earliest days, when Schrödinger's papers were appearing in the *Annalen der Physik*, Fermi explained them to his students in private seminars; later he recast some of Dirac's papers in more familiar form, in part for didactical reasons. As time went on, his treatment and his courses became more systematic; there must be a number of notebooks of his lectures as recorded by students at the University of Rome, at Columbia University, and at the University of Chicago.

Early in 1954, less than a year before his untimely death, Fermi again gave a course in quantum mechanics at the University of Chicago. This time, however, he prepared the notes for the students himself by writing the outlines of the lectures on duplicator master sheets and delivering copies to the students in advance of each lecture.

The publisher, advised by friends and former pupils of Fermi, has decided to reproduce these lecture outlines in an inexpensive edition, in order to make them available to a larger group of students than those who had the privilege of attending the original lectures.

We hope that young physicists of the coming generation who have never come in direct contact with Fermi, and for whom he must be little more than a name among the great scientists of the century, will enjoy having a notebook on such an important topic as quantum mechanics written for them by such a master in his own hand.

Having pointed out the genesis of these notes, we do not need to emphasize that they cannot be construed in any way as the final presentation of quantum mechanics by Fermi, such as he could have given in a more elaborate text. Heisenberg, Pauli, Dirac, de Broglie, Jordan, Kramers, to mention only some of the creators of quantum mechanics, have all presented their own versions of quantum mechanics in books which are justly famous. The notes by Fermi are not to be compared in any way with these texts. They are written in a spirit and for a purpose completely different from that of the works mentioned above.

Fermi in the last ten or fifteen years of his life scarcely read any book on physics. He kept abreast of scientific developments mainly by hearing the results of investigations and reconstructing them on his own. It is practically certain that he did not consult any text of quantum mechanics while compiling these notes, except perhaps in a very minor fashion. If sections of the notes are very similar to some standard treatments, we must assume that, in rethinking the subject, Fermi arrived in his own way at the formulations contained in these notes.

We repeat that the notes were clearly prepared only for the lectures and that their distribution beyond the class group was not intended by the author. It is only because we know his great interest in teaching that we think it is not irreverent to his memory to publish the notes for the benefit of other students.

E. SEGRÈ

BERKELEY, CALIFORNIA

Quantum Mechanics

E. Fermi Physics 341
Winter 1954

1 - Optics - Mechanics analogy:

Dictionary

Mass point	Wave packett
Trajectory	Ray
Velocity (V)	Group velocity (V)
No simple analog	Phase velocity (v)
Potential function of position $U(x)$	Refractive index (or v) function of position
(1) Energy (W) $W = W(\nu)$	Frequency (ν) (dispersive medium $v(\nu, x)$)

First: Trajectory = Ray

↓ from Maupertuis ↓ from Fermat

$$(2) \quad \int \sqrt{W-U}\, ds = \min; \quad \int \frac{ds}{v} = \min \quad (3)$$

Proof of Maupertuis:

$$\delta \int \sqrt{W-U}\, ds = \int \left(\sqrt{W-U}\, \delta ds - \frac{\delta U}{2\sqrt{W-U}} ds \right) = 0$$

use $\delta ds = \sum \frac{dx}{ds} \delta dx$, $\delta U = \sum \frac{\partial U}{\partial x} \delta x$

and part. integr., Find minimum equations

$$\frac{d}{ds}\left(\sqrt{W-U} \frac{dx}{ds} \right) = -\frac{1}{2\sqrt{W-U}} \frac{\partial U}{\partial x}$$

use $V = \sqrt{\frac{2}{m}} \sqrt{W-U}$, $dt = \frac{ds}{V} = \sqrt{\frac{m}{2}} \frac{ds}{\sqrt{W-U}}$

$$\rightarrow m \frac{d^2x}{dt^2} = -\frac{\partial U}{\partial x}$$

Therefore: (2) is true because of eq. of motion

Proof of Fermat

$\int \frac{ds}{v} = \min \rightarrow \nu \int \frac{ds}{v} = \min \rightarrow \int \frac{ds}{\lambda} = \min \rightarrow$ No of waves = min

means; no of waves stationary: hence positive interference.

From (1) (2) Trajectory – Ray if

(4) $$\frac{1}{v(\nu, x)} = f(\nu)\sqrt{W(\nu) - U(x)}$$

$f(\nu)$ and $W = W(\nu)$ so far arb. fcts

Determine f & W from:

Vel. of mass pt $V = \sqrt{\frac{2}{m}}\sqrt{W - U}$ equals

Group vel. of pckt $V = 1 \Big/ \frac{d}{d\nu}\left(\frac{\nu}{v}\right)$

Proof of group vel. formula

Wave packett with small frequency spread

$$\sum a \cos 2\pi\nu\left(t - \frac{x}{v(\nu)}\right)$$

If all a's > 0 constructive interf at $x = 0$ and $t = 0$. Locate now packett for $t \neq 0$ by demanding constructive interference.

Required $$\frac{d}{d\nu}\left\{\nu\left(t - \frac{x}{v(\nu)}\right)\right\} = 0$$

or $$t = x\frac{d}{d\nu}\frac{\nu}{v}$$ identify this to $t = \frac{x}{V}$

Find

(5) $$\boxed{\frac{1}{V} = \frac{d}{d\nu}\frac{\nu}{v(\nu)}}$$

Condition becomes

(6) $$\frac{d}{d\nu}\frac{\nu}{v} = \sqrt{\frac{m}{2}}\frac{1}{\sqrt{W(\nu) - U}}$$

Use (4)

$$\sqrt{\frac{m}{2}}\frac{1}{\sqrt{W - U}} = \frac{d}{d\nu}\left\{\nu f\sqrt{W(\nu) - U}\right\} = \frac{d(\nu f)}{d\nu}\sqrt{W - U} + \frac{\nu f}{12}\frac{dW/d\nu}{\sqrt{W - U}}$$

U varies from place to place indep. of ν therefore $\sqrt{W-U}$ is cons. as indep. Find then conditions:

$$\frac{d(\nu f)}{d\nu} = 0 \qquad \sqrt{\frac{m}{2}} = \frac{\nu f}{2}\frac{dW}{d\nu}$$

$\nu f = \text{constant}$

$\frac{dW}{d\nu} = \text{constant} = h$

$W = h\nu + \text{const} = h\nu$

set this $= 0$ by suitable choice of energy constant

Therefore result

$$(7)\quad W = h\nu$$

$$(8)\quad f = \frac{\sqrt{2m}}{h\nu}$$

$$(9)\quad v = \frac{h\nu}{\sqrt{2m}}\frac{1}{\sqrt{h\nu - U}}$$ determines refractive index and dispersion everywhere

Change to angular frequency

$$(10)\qquad \omega = 2\pi\nu \qquad \text{also put} \quad \hbar = h/2\pi$$

Final result

$$W = \hbar\omega \qquad v = \frac{\hbar\omega}{\sqrt{2m}}\frac{1}{\sqrt{\hbar\omega - U}} \qquad V = \sqrt{\frac{2}{m}}\sqrt{\hbar\omega - U}$$

$$(11)\quad \lambda\!\!\!^{-} = \frac{\lambda}{2\pi} = \frac{v}{\omega} = \frac{\hbar}{\sqrt{2m}}\frac{1}{\sqrt{\hbar\omega - U}} = \frac{\hbar}{mV} = \frac{\hbar}{p}$$

(de Broglie wave length)

Experiments on material particle diffraction may be used to determine λ hence h or $\hbar$

$h = 6.6252(5) \times 10^{-27}$ ergs sec (L^2MT^{-1})

$\hbar = 1.05444(9) \times 10^{-27}$ "

2 – Schroedinger equation.

(1) $v = v(\omega, P) = \frac{\hbar\omega}{\sqrt{2m}}\frac{1}{\sqrt{\hbar\omega - U}}$

Monochromatic wave equation

$$\nabla^2\psi - \frac{1}{v^2}\frac{\partial^2\psi}{\partial t^2} = 0$$

((comments: need to assume fixed ω))

(2) $\psi = u\, e^{-i\omega t} = u\, e^{-\frac{i}{\hbar}Wt}$

$$\nabla^2 u + \frac{\omega^2}{v^2}u = 0 \qquad \nabla^2 u + \frac{2m}{\hbar^2}(\hbar\omega - U)u = 0$$

write $\omega u \sim -\frac{1}{i}\frac{\partial\psi}{\partial t}$

Time dependent Schrodinger equation

(3) $\nabla^2\psi + \frac{2mi}{\hbar}\frac{\partial\psi}{\partial t} - \frac{2m}{\hbar^2}U\psi = 0$

Written also as

(4) $i\hbar\frac{\partial\psi}{\partial t} = -\frac{\hbar^2}{2m}\nabla^2\psi + U\psi$

(Comments: ψ complex)

Time dep. equation (assuming (2))

(5) $W\psi = -\frac{\hbar^2}{2m}\nabla^2\psi + U\psi$

Valid only for states of fixed energy $W = \hbar\omega$

Continuity equation for (4)

Write conjugate equation

(6) $-i\hbar\frac{\partial\psi^*}{\partial t} = -\frac{\hbar^2}{2m}\nabla^2\psi^* + U\psi^*$

(4) $\times\, \psi^*$ – (6) $\times\, \psi$ yields

(7) $\frac{\partial}{\partial t}(\psi^*\psi) + \nabla\cdot\left\{\frac{\hbar}{2mi}(\psi^*\nabla\psi - \psi\nabla\psi^*)\right\}$

Suggested provisional interpretation

(8) $$\psi^*\psi = |\psi|^2 = \text{density of probability}$$

(9) $$\frac{\hbar}{2mi}\left(\psi^*\nabla\psi - \psi\nabla\psi^*\right) = \text{average value of flow density}$$

Normalization: (8) suggests to determine ψ such that

(10) $$\int|\psi|^2 d\tau = \int\psi^*\psi\, d\tau = 1$$

This requires certain conditions

a) Near singular pt ψ less ∞ than $r^{-3/2}$

b) Limit of infinite distance $\psi \to 0$ faster than $r^{-3/2}$

Exceptions to rule (b) will have to be considered later

Generalizations.

Point on line

(11) $$\begin{cases} i\hbar\dfrac{\partial\psi}{\partial t} = -\dfrac{\hbar^2}{2m}\dfrac{\partial^2\psi}{\partial t^2} + U(x)\psi \\ \text{or} \\ E\,u(x) = -\dfrac{\hbar^2}{2m}\dfrac{d^2u}{dx^2} + U(x)u \end{cases}$$

Rotator with fixed axis

A = mom. of inertia

(12) $$\begin{cases} i\hbar\dfrac{\partial\psi}{\partial t} = -\dfrac{\hbar^2}{2A}\dfrac{\partial^2\psi}{\partial\alpha^2} + U(x)\,\psi(\alpha,t) \\ \text{or} \\ E\,u(\alpha) = -\dfrac{\hbar^2}{2A}\dfrac{d^2u}{d\alpha^2} + U(\alpha)\,u(\alpha) \end{cases}$$

Point on sphere or dumbbell with fixed c. of grav.

(13) $$\Lambda\psi = \frac{1}{\sin\vartheta}\frac{\partial}{\partial\vartheta}\left(\sin\vartheta\frac{\partial\psi}{\partial\vartheta}\right) + \frac{1}{\sin^2\vartheta}\frac{\partial^2\psi}{\partial\varphi^2}$$

$$(14)\begin{cases}\Lambda\psi - \frac{2A}{\hbar^2}U(\vartheta,\varphi)\psi = -\frac{2Ai}{\hbar}\frac{\partial\psi}{\partial t}\\ \Lambda u + \frac{2A}{\hbar^2}(E-U)u = 0\end{cases}$$

$A = \begin{cases} r^2 m \text{ or} \\ \text{mom of} \\ \text{inertia}\end{cases}$

Several mass points

$$\psi(t, x_1, y_1, z_1, x_2, y_2, z_2, \ldots, x_n, y_n, z_n)$$

$$(15)\begin{cases} i\hbar\frac{\partial\psi}{\partial t} = -\frac{\hbar^2}{2}\sum_1^n \frac{1}{m_j}\nabla_j^2\psi + U\psi \\ Eu = -\frac{\hbar^2}{2}\sum_j \frac{1}{m_j}\nabla_j^2 u + Uu\end{cases}$$

General dynamical system

Sum over equal indices

$$(16)\quad T = \frac{1}{2}m_{ik}\dot{q}_i\dot{q}_k$$

Define $m^{ik}m_{il} = \delta_{kl}$

$m^{il} = \frac{\text{minor of } m_{il}}{D}$

$$(17)\quad D = \det|m_{ik}|$$

$$(18)\quad \nabla^2\psi = \frac{1}{\sqrt{D}}\frac{\partial}{\partial q_k}\left(\sqrt{D}\, m^{kl}\frac{\partial\psi}{\partial q_l}\right)$$

(19) Volume element

$$d\tau = \sqrt{D}\, dq_1 dq_2 \ldots dq_n$$

Equation

$$(20)\begin{cases} -\frac{\hbar^2}{2}\nabla^2\psi + U\psi = i\hbar\frac{\partial\psi}{\partial t} \\ -\frac{\hbar^2}{2}\nabla^2 u + Uu = Eu\end{cases}$$

3 – Simple one dimensional problems

Time indep. equation

(1) $$u'' + \frac{2m}{\hbar^2}(E - U)\,u = 0$$

a) Closed line, length a, $V(x) = 0$

(2) $$u \sim e^{\pm i\sqrt{\frac{2mE}{\hbar^2}}\,x}$$

Periodicity condition requires $u \sim e^{\frac{2\pi i}{a}\ell x}$

Therefore $\ell =$ integer

(3) $$E_\ell = \frac{2\pi^2\hbar^2}{ma^2}\ell^2$$

Comments on quantization of energy

Normalized functions

(4) $$u_\ell = \frac{1}{\sqrt{a}}\,e^{\frac{2\pi i\ell}{a}x}$$

b) Rotator with fixed axis. As above with
$m \to A =$ mom. of inertia
$a \to 2\pi$
$x \to \alpha$

(5) $$\begin{cases} E_\ell = \dfrac{\hbar^2}{2A}\ell^2 \\ u_\ell = \dfrac{1}{\sqrt{2\pi}}\,e^{i\ell\alpha} \end{cases}$$

c) Boundary condition where $U = \infty$

$V(x)$, ∞, x

Inside wall

$$u \sim e^{-\sqrt{\frac{2mU}{\hbar^2}}\,x}$$

(reject e^{+} solution because too infinite on right)

at wall $$\frac{u'}{u} = -\sqrt{\frac{2mU}{\hbar^2}} \to \infty$$

(6) Therefore: at wall take $\begin{cases} u = 0 \\ u' \text{ finite} \end{cases}$

d) Point on segment (from $x=0$ to $x=a$)

Potential $=0$ on segment, becomes ∞ at ends

Therefore $u(0)=u(a)=0$ are boundary conditions

Solution of

$$u'' + \frac{2mE}{\hbar^2}u = 0$$

$$u \sim \begin{matrix}\sin \\ \cos\end{matrix}\sqrt{\frac{2mE}{\hbar^2}}x$$ (because of $u(0)=0$ reject cosine)

$$u \sim \sin\sqrt{\frac{2mE}{\hbar^2}}x$$ Because of $u(a)=0$

must be

$$\sqrt{\frac{2mE}{\hbar^2}}\,a = n\pi \quad (n \text{ integer})$$

Therefore

$$(7)\quad \begin{cases} E_n = \dfrac{\pi^2\hbar^2}{2a^2m}n^2 \\ u_n = \sqrt{\dfrac{2}{a}}\sin\dfrac{\pi n x}{a} \end{cases}$$

normalization factor

e) Point on infinite line – Zero potential

$$(8)\quad u'' + \frac{2mE}{\hbar^2}u = 0$$

has solutions

$$(9)\quad e^{\pm i\sqrt{\frac{2mE}{\hbar^2}}x}$$

None of these is normalizable!

Get around difficulty in two ways:

1 – As limit of case a)

$$u_l = \frac{1}{\sqrt{a}}e^{\frac{2\pi i l}{a}x} \qquad a \to \infty$$

$$E_l = \frac{2\pi^2\hbar^2}{m}\left(\frac{l}{a}\right)^2$$

Energy levels are quasi-continuous

No of levels in dE is obtained from

$$\frac{dE}{dl} = \frac{4\pi^2\hbar^2}{a^2 m} l = \frac{2\pi\hbar}{a}\sqrt{\frac{2}{m}}\sqrt{E}$$

$$\text{No of levels} = \frac{2}{dE/dl} dE = \frac{a}{\pi\hbar}\sqrt{\frac{m}{2}} \frac{dE}{\sqrt{E}}$$

factor 2 because l may be pos. or negative

(becomes ∞ for $a \to \infty$)

In limit: Continuous spectrum with all values $E \geq 0$ allowable

Note. Same result could be found by limit $a \to \infty$ in case d)

Alternate approach: Sharp energy levels do not exist but wave packetts like

$$u_{\delta k} = \int_{k_0 - \frac{\delta k}{2}}^{k_0 + \frac{\delta k}{2}} e^{ikx} dk = \frac{2}{x} \sin \frac{x \delta k}{2} \; e^{ik_0 x}$$

are normalizable for δk very small. They correspond to almost definite energy.

More on this later with uncertainty principle

4 – Linear oscillator

(1) $$U = \frac{m}{2}\omega^2 x^2$$

Schroedinger eq.

(2) $$u'' + \frac{2m}{\hbar^2}\left(E - \frac{m\omega^2}{2}x^2\right)u = 0$$

Put

(3) $$\xi = \sqrt{\frac{m\omega}{\hbar}}\,x \qquad \varepsilon = \frac{2E}{\hbar\omega}$$

(4) $$\frac{d^2u}{d\xi^2} + (\varepsilon - \xi^2)\,u = 0$$

(5) $$u = v(\xi)\,e^{-\xi^2/2}$$

(6) $$\frac{d^2v}{d\xi^2} - 2\xi\frac{dv}{d\xi} + (\varepsilon - 1)\,v = 0$$

Series exp.

(7) $$v = \sum a_r \xi^r \quad \text{yields}$$

(8) $$a_{r+2} = \frac{2r+1-\varepsilon}{(r+1)(r+2)}\,a_r$$

r even and r odd yield two indep. solutions. $v(\infty) \rightarrow e^{\xi^2}$ (not allowable) except for

(9) $$\varepsilon = 2n+1$$

Then either even or odd solution is a polynomial (Hermite)

(10) $$\begin{cases} H_0(\xi) = 1 \quad H_1(\xi) = 2\xi \quad H_2(\xi) = -2 + 4\xi^2 \\ H_3(\xi) = -12\xi + 8\xi^3 \end{cases}$$

general expression:

(11) $$H_n(\xi) = (-1)^n e^{\xi^2}\frac{d^n}{d\xi^n}e^{-\xi^2}$$

Proof: (5), that is

(12) $H_n'' - 2\xi H_n' + 2n H_n = 0$

is equivalent (11) to

(13) $\left\{ \frac{d^{n+2}}{d\xi^{n+2}} + 2\xi \frac{d^{n+1}}{d\xi^{n+1}} + (2+2n)\frac{d^n}{d\xi^n} \right\} e^{-\xi^2} = 0$

Verify for $n=0$; then by successive derivation for $n = 1, 2, \ldots$

Useful properties

(14) $\frac{dH_n}{d\xi} = 2n H_{n-1}(\xi)$

(Proof: equivalent to (13) written for $n-1$)

Normalization property:

(15) $\int_{-\infty}^{\infty} H_n^2(\xi)\, e^{-\xi^2} d\xi = \sqrt{\pi}\, 2^n n!$

[Proof: By induction – First directly for $n=0$ Then use (11) & (14) to proove induction property $\int_{-\infty}^{\infty} H_n^2 e^{-\xi^2} d\xi = 2n \int_{-\infty}^{\infty} H_{n-1}^2 e^{-\xi^2} d\xi$]

Integral property

(16) $\int_{-\infty}^{\infty} H_n(x)\, e^{-x^2} e^{ipx} dx = i^m \sqrt{\pi}\, p^m e^{-p^2/4}$

[Proof: directly for $n=0$; then by induction with (11)]

Normalized oscillator eigenfunctions

(17) $u_n = \left(\frac{m\omega}{\hbar}\right)^{1/4} \frac{1}{\sqrt{\sqrt{\pi}\, 2^n n!}} H_n(\xi)\, e^{-\xi^2/2}$ $\xi = \sqrt{\frac{m\omega}{\hbar}}\, x$

(18) $E_n = \hbar\omega(n + \tfrac{1}{2})$ (Comments)

5 – WKB method

(1) $u'' + \frac{2m}{\hbar^2}(E - V(x))\, u = 0$

$g = \frac{2m}{\hbar^2}(E-U) = \frac{m^2V^2}{\hbar^2}$

V = class. velocity

(2) $u'' + g(x)\, u = 0$

Assume first $g(x) > 0$

(3) $u = e^{iy(x)}$ into (2)

(4) $y'^2 - iy'' = g$ First guess:

$$y' \approx \sqrt{g} \quad \text{then} \quad \frac{y''}{y'^2} = \frac{g'}{2g^{3/2}}$$

Therefore: guess is fair approximation when

(5) $$|g'| \ll 2g^{3/2}$$

Put then

(6) $y' = \sqrt{g} + \varepsilon$

(Neglect ε^2 and ε' or ε'' terms to find)

$$g + 2\varepsilon\sqrt{g} - \frac{ig'}{2\sqrt{g}} = g \longrightarrow \varepsilon = \frac{ig'}{4g}$$

(7) $$y \approx \int\left(\sqrt{g} + \frac{ig'}{4g}\right) dx = \int \sqrt{g}\, dx + \frac{i}{4}\log g$$

(8) $$u = e^{iy} \approx \frac{1}{g^{1/4}} e^{i\int\sqrt{g}\,dx}$$

Other solutions → $\frac{1}{g^{1/4}} e^{-i\int\sqrt{g}\,dx}$ or real linear combination

(9) $$u \sim \frac{1}{g^{1/4}} \sin\left\{\int\sqrt{g}\, dx + \text{const}\right\}$$

[Note: $|u|^2 \sim \frac{1}{\sqrt{g}} \sim \frac{1}{V} \sim$ time classically spent at location x].

Case $g(x) < 0$

Find similarly

(10) $$u \sim \frac{1}{(-g)^{1/4}} e^{\pm \int \sqrt{-g(x)}\, dx} \quad \text{for } g<0$$

$u \approx e^{\int \sqrt{-g}\, dx}$ $\quad$ $u \approx \sin\left(\int \sqrt{g}\, dx + \text{const}\right)$ $\quad$ $u \approx e^{-\int \sqrt{-g}\, dx}$

$\leftarrow g<0 \rightarrow$ $\quad$ $\leftarrow g>0 \rightarrow$ $\quad$ $\leftarrow g<0 \rightarrow$

V, E, x

Matching of solutions where g changes sign

Equation

(11) $$\omega'' + x\omega = 0$$

has solution

(12) $$\omega = \sqrt{x}\left\{ c_1 J_{1/3}\left(\tfrac{2}{3}x^{2/3}\right) + c_2 N_{1/3}\left(\tfrac{2}{3}x^{2/3}\right)\right.$$

Linear comb. that vanishes at $-\infty$ has asymptotic expressions,

(13) $$\omega(x) \begin{cases} \dfrac{1}{x^{1/4}} \sin\left(\tfrac{2}{3}x^{3/2} + \tfrac{\pi}{4}\right) & x \to \infty \\ \dfrac{1}{2(-x)^{1/4}} e^{-\frac{2}{3}(-x)^{3/2}} & x \to -\infty \end{cases}$$

Compare with WKB solution.

Conclusion: at each end points of interval where $g>0$ add phase $\pi/4$

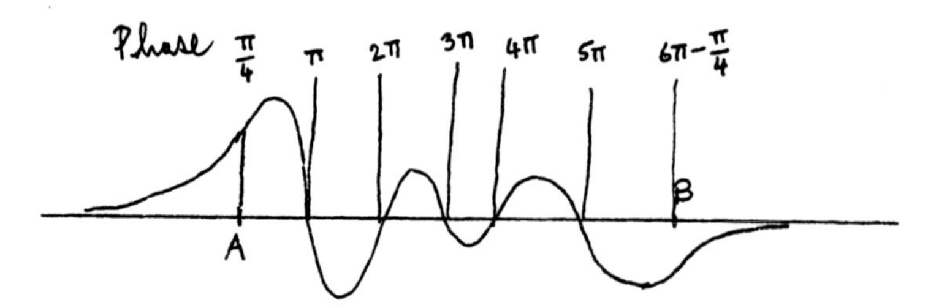

Let $g > 0$ between A, B and $g < 0$ outside AB

Phase difference B to A

$$(n + \tfrac{1}{2})\pi$$

n = number of nodes between A + B.

Condition for matching from A to B

$$(n+\tfrac{1}{2})\pi = \int_A^B \sqrt{g}\, dx = \int_A^B \frac{mV}{\hbar}\, dx =$$

($p = mV$ = classical momentum) $= \frac{1}{2\hbar}\oint p\, dx$

Conclusion, Bohr, Sommerfeld quantization condition

(14) $$\oint p\, dx = 2\pi\hbar\left(n+\tfrac{1}{2}\right)$$

Note: Slightly different conditions on completely accessible closed path

(15) $$\oint p\, dx = 2\pi\hbar\, n$$

Or on completely accessible segment bounded by infinitely hy potential walls at A and B

(16) $$\oint p\, dx = 2\pi\hbar(n+1)$$ n = no of nodes inside segment

6 – Spherical harmonics

Legendre polinomials

$$(1)\qquad P_\ell(x) = \frac{1}{2^\ell \ell!}\frac{d^\ell}{dx^\ell}(x^2-1)^\ell$$

$$(2)\qquad (1-x^2)P_\ell'' - 2xP_\ell' + \ell(\ell+1)P_\ell = 0$$

$$(3)\qquad \int_{-1}^{1} P_\ell^2(x)\,dx = \frac{2}{2\ell+1}$$

$$(4)\qquad \int_{-1}^{1} P_\ell(x)P_{\ell'}(x)\,dx = 0 \quad \text{for } \ell \neq \ell'$$

$$(5)\qquad P_\ell = \frac{2\ell-1}{\ell}\,x\,P_{\ell-1} - \frac{\ell-1}{2}P_{\ell-2}$$

$$(6)\quad \begin{cases} P_0 = 1 \qquad P_1 = x \qquad P_2 = \frac{3}{2}x^2 - \frac{1}{2} \\ P_3 = \frac{5}{2}x^3 - \frac{3}{2}x \qquad P_4 = \frac{35}{8}x^4 - \frac{15}{4}x^2 + \frac{3}{8} \\ P_5 = \frac{63}{8}x^5 - \frac{35}{4}x^3 + \frac{15}{8}x \;;\; P_\ell(1) = 1 \end{cases}$$

Alternate definition

$$(7)\qquad \frac{1}{\sqrt{1-2rx+r^2}} = \sum_0^\infty P_\ell(x)\,r^\ell$$

Spherical harmonics:

$$(8)\quad \begin{cases} Y_{\ell m}(\vartheta,\varphi) = \frac{1}{N_{\ell m}} e^{im\varphi} \sin^{|m|}\vartheta \,\frac{d^{|m|} P_\ell(\cos\vartheta)}{d(\cos\vartheta)^{|m|}} \\ \frac{1}{N_{\ell m}} = \pm\frac{1}{\sqrt{2\pi}}\sqrt{\frac{2\ell+1}{2}\frac{(\ell-|m|)!}{(\ell+|m|)!}} \quad \begin{matrix}\text{for } m\le 0 \ + \text{sign} \\ \text{for } m>0\ (-1)^m \text{ sign}\end{matrix} \end{cases}$$

Normalization

$$(9)\quad \int_{4\pi} Y^*_{lm} Y_{l'm'}\, d\omega = \delta_{ll'}\delta_{mm'}$$

Diff. equation

$$(10)\quad \Lambda Y_{lm} + l(l+1)Y_{lm} = 0$$

$$(11)\quad \Lambda = \frac{1}{\sin\vartheta}\frac{\partial}{\partial\vartheta}\left(\sin\vartheta\frac{\partial}{\partial\vartheta}\right) + \frac{1}{\sin^2\vartheta}\frac{\partial^2}{\partial\varphi^2}$$

$$(12)\quad \begin{cases} \nabla^2\left(r^l Y_l\right) = 0 \\ \nabla^2\left(r^{-l-1} Y_l\right) = 0 \quad \text{(except origin)} \end{cases}$$

$$(13)\quad \nabla^2 = \frac{\partial^2}{\partial r^2} + \frac{2}{r}\frac{\partial}{\partial r} + \frac{1}{r^2}\Lambda$$

Development in sph. harm.:

$$(14)\quad \begin{cases} f(\vartheta,\varphi) = \sum c_{lm} Y_{lm}(\vartheta,\varphi) \\ c_{lm} = \int_{4\pi} f Y^*_{lm}\, d\omega \end{cases}$$

$$Y_{00} = 1/\sqrt{4\pi} \qquad Y_{10} = \sqrt{\frac{3}{4\pi}}\cos\vartheta$$

$$Y_{1,\pm1} = \mp\sqrt{\frac{3}{8\pi}}\sin\vartheta\, e^{\pm i\varphi}$$

$$Y_{20} = \sqrt{\frac{5}{4\pi}}\left(\frac{3}{2}\cos^2\vartheta - \frac{1}{2}\right) \qquad Y_{2,\pm1} = \mp\sqrt{\frac{15}{8\pi}}\sin\vartheta\cos\vartheta\, e^{\pm i\varphi}$$

$$Y_{2,\pm2} = \frac{1}{4}\sqrt{\frac{15}{2\pi}}\sin^2\vartheta\, e^{\pm 2i\varphi}$$

$$Y_{30} = \sqrt{\frac{7}{4\pi}}\left(\frac{5}{2}\cos^3\vartheta - \frac{3}{2}\cos\vartheta\right)$$

$$Y_{3,\pm1} = \mp\frac{1}{4}\sqrt{\frac{21}{4\pi}}\sin\vartheta\left(5\cos^2\vartheta - 1\right)e^{\pm i\varphi}$$

$$Y_{3,\pm2} = \frac{1}{4}\sqrt{\frac{105}{2\pi}}\sin^2\vartheta\cos\vartheta\, e^{\pm 2i\varphi}$$

$$Y_{3,\pm3} = \mp\frac{1}{4}\sqrt{\frac{35}{4\pi}}\sin^3\vartheta\, e^{\pm 3i\varphi}$$

7 - Central forces

(1) $\nabla^2 u + \frac{2m}{\hbar^2}\left(E - U(r)\right) u = 0$

Polar coordinates

(2) $\frac{\partial^2 u}{\partial r^2} + \frac{2}{r}\frac{\partial u}{\partial r} + \frac{1}{r^2}\Lambda u + \frac{2m}{\hbar^2}\left(E - U(r)\right) u = 0$

Develop $u(r, \vartheta, \varphi)$ in sph. harm.

(3) $u = \sum R_{lm}(r)\, Y_{lm}(\vartheta, \varphi)$

Use (6-10)

(4) $\sum Y_{lm}\left\{ R''_{lm} + \frac{2}{r} R'_{lm} - \frac{l(l+1)}{r^2} R_{lm} + \frac{2m}{\hbar^2}(E-U) R_{lm} \right\} = 0$

Multiply by $Y^*_{lm}\, d\omega$ and integrate. Find

(5) $R''_l + \frac{2}{r} R'_l + \frac{2m}{\hbar^2}\left\{ E - U(r) - \frac{\hbar^2}{2m}\frac{l(l+1)}{r^2} \right\} R_l = 0$

Note: indep. of m.

Each solution of (5) yields $2l+1$ solutions of (1)

Useful transformation

(6) $r R_l(r) = v_l(r)$

(7) $v''_l(r) + \frac{2m}{\hbar^2}\left\{ E - U(r) - \frac{\hbar^2}{2m}\frac{l(l+1)}{r^2} \right\} v_l(r) = 0$

$l=0$	$l=1$	$l=2$	$l=3$	$l=4$	$l=5$	$l=6$
s	p	d	f	g	h	i

Will prove later $\hbar l \sim$ ang momentum $= M$

Two mass points, central forces

(8) $$+\frac{1}{m_1}\nabla_1^2 u + \frac{1}{m_2}\nabla_2^2 u + \frac{2}{\hbar^2}\left(E - U(r)\right)u = 0$$

Change coordinates

(9) $$\begin{cases} x = x_2 - x_1 & \text{(relative coordinates)} \\ X = \dfrac{m_1 x_1 + m_2 x_2}{m_1 + m_2} & \text{(c. of mass coordinates)} \end{cases}$$

Also

$$\nabla^2 = \frac{\partial^2}{\partial x^2} + \dots \qquad \nabla_g^2 = \frac{\partial^2}{\partial X^2} + \dots$$

(10) $$\begin{cases} \dfrac{1}{m_1}\nabla_1^2 + \dfrac{1}{m_2}\nabla_2^2 = \dfrac{1}{m_1 + m_2}\nabla_g^2 + \dfrac{1}{m}\nabla^2 \\ m = \dfrac{m_1 m_2}{m_1 + m_2} = \text{red. mass} \end{cases}$$

(8) becomes:

(11) $$\frac{1}{m_1 + m_2}\nabla_g^2 u + \frac{1}{m}\nabla^2 u + \frac{2}{\hbar^2}\left(E - U(r)\right)u = 0$$

(12) $$u(x, X) = \sum w_k(x, y, z)\, e^{i\vec{k}\cdot\vec{X}}$$

Substitute and invert Fourier

(13) $$\nabla^2 w_k + \frac{2m}{\hbar^2}\left(E_{rel} - U(r)\right)w_k = 0$$

(14) $$E_{rel} = E - \frac{(\hbar k)^2}{2(m_1 + m_2)}$$

energy of c. of mass motion

Conclusion: Separation of relative and c. of m. motion like in class. mech.!

8-

Hydrogen Atom

(1) $U = -\frac{Ze^2}{r}$ (Neglect nuclear motion; m will be reduced mass)

Radial equation (7-7)

(2) $$v''(r) + \frac{2m}{\hbar^2}\left(E + \frac{Ze^2}{r} - \frac{\hbar^2}{2m}\frac{\ell(\ell+1)}{r^2}\right)v(r) = 0$$

Put

(3) $$\begin{cases} x = 2r/r_0 & r_0 = \sqrt{\dfrac{\hbar^2}{2m|E|}} \\ A = \dfrac{Ze^2}{2r_0|E|} = \sqrt{\dfrac{mZ^2e^4}{2\hbar^2|E|}} \end{cases}$$

(4) $$\frac{d^2v}{dx^2} + \underbrace{\left(\pm\frac{1}{4} + \frac{A}{x} - \frac{\ell(\ell+1)}{x^2}\right)}_{g(x)}v = 0$$ (+ for $E>0$; − for $E<0$)

Graphical discussion

$g(x)$, x

$E<0$: $v \to e^{-\frac{1}{2}x}$ and not $e^{+\frac{1}{2}x}$ — therefore: adjustment required. Only discreet values of E allowable

$E>0$: $v \to \begin{matrix}\sin \\ \text{or} \\ \cos\end{matrix}\left(\frac{x}{2}\right)$ — no condition needed at $x\to\infty$. All $E>0$ allowable

Assume $E<0$ — Case of discreet e.values.

(5) $$\frac{d^2v}{dx^2} + \left(-\frac{1}{4} + \frac{A}{x} - \frac{\ell(\ell+1)}{x^2}\right)v = 0$$

(6) $$v(x) = e^{-x/2}\,y(x)$$

(7) $$y'' - y' + \left(\frac{A}{x} - \frac{\ell(\ell+1)}{x^2}\right)y = 0$$

$$y(x\to 0) = \begin{cases} x^{\ell+1} \text{ or} \\ x^{-\ell} \end{cases}$$

$y \to x^{-\ell}$ corresp. to $u \sim r^{-\ell-1}$. Normalization divergent at origin for $\ell \geqslant 1$. Therefore reject. For $\ell=0$ also reject because $u \sim 1/r$ and $\nabla^2 \frac{1}{r} = -4\pi\delta(\vec{r})$ But no such singularity in potential!

Therefore acceptable solution

(8) $$y(x) = x^{\ell+1} \sum_0^\infty a_s x^s$$

Substitute in (7). Find

(9) $$a_{s+1} = \frac{s+\ell+1-A}{(s+1)(s+2\ell+2)} a_s$$

In general infinite series — This too large at infinite ($y(x\to\infty) \sim e^x$; $u \to e^{x/2}$ un normalizable,

Only acceptable solutions when A = int. number

(10) $$A = n = n' + \ell + 1$$

Then series $\to$ polynomial!

$R_\infty = \frac{me^4}{2\hbar^2} = 21.795 \times 10^{-12}$ erg $= 13.605$ eV $= 109737.309(12)$ cm^{-1}

(10) + (3) give

(11) $$E_n = -\frac{mZ^2e^4}{2\hbar^2 n^2} \qquad n = \ell+1, \ell+2, \ldots$$

Solution expressible in Laguerre Polynomials

(12) $$L_k(x) = e^x \frac{d^k}{dx^k}\left(x^k e^{-k}\right)$$

(13) $$\begin{cases} L_0 = 1 \quad L_1 = 1-x \quad L_2 = 2-4x+x^2 \\ L_3 = 6-18x+9x^2-x^3 \end{cases}$$

Put

$$f(x) = x^k e^{-x}$$

$$L_k = e^x f^{(k)}(x)$$

$$x f' = (k - x) f$$

Diff. $(k+1)$ times

$$x f^{(k+2)} + (x+1) f^{(k+1)} + (k+1) f^{(k)} = 0$$

$$f^{(k)} = e^{-x} L_k$$ yields

(14) $$x L_k'' + (1-x) L_k' + k L_k = 0$$

This is Laguerre diff. equation

(15) $$L_k^{(j)}(x) = \frac{d^j}{dx^j}\left\{e^x \frac{d^k}{dx^k}(x^k e^{-x})\right\}$$

$\frac{d^j}{dx^j}$ (14) →

(16) $$x L_k^{(j)''} + (j+1-x) L_k^{(j)'} + (k-j) L_k^{(j)} = 0$$

~~Orthogonality~~ Normalization property

(17) $$\int_0^\infty L_k^{(j)} L_{k'}^{(j)} x^j e^{-x} dx = \frac{(k!)^3}{(k-j)!}\delta_{kk'}$$

Normalized e.f's

(18) $$\begin{cases} u_{nlm} = R_{nl}(r) Y_{lm}(\vartheta, \varphi) \\ R_{nl} = \sqrt{\dfrac{4(n-l-1)!}{a^3 n^4 [(n+l)!]^3}}\, e^{-\frac{r}{na}} \left(\dfrac{2r}{na}\right)^l L_{n+l}^{(2l+1)}\left(\dfrac{2r}{na}\right) \end{cases}$$

(19) $$a = \frac{\hbar^2}{me^2}\frac{1}{Z}$$ $$\frac{\hbar^2}{me^2} = \text{Bohr radius (nucleus of infinite mass)} = 0.529171(6)\times 10^{-8}\text{ cm.}$$

(20)
$$\begin{cases} u(1s) = \dfrac{1}{\sqrt{\pi a^3}}\, e^{-r/a} \\[2ex] u(2s) = \dfrac{(2 - r/a)\, e^{-r/2a}}{4\sqrt{2\pi a^3}} \\[2ex] u(2p) = \dfrac{\frac{r}{a}\, e^{-\frac{r}{2a}}}{8\sqrt{\pi a^3}} \begin{cases} -\sin\vartheta\, e^{i\varphi} \\ \sqrt{2}\cos\vartheta \\ \sin\vartheta\, e^{-i\varphi} \end{cases} \end{cases}$$

Note: s-wave functions are the only ones for which $u(r=0) \neq 0$. For them

(21) $$u_{ns}(r=0) = \frac{1}{\sqrt{\pi a^3 n^3}}$$

} cont. spectrum

$-R/9$ 3s, 3p

$-R/4$ 2s, 2p

$-R$ 1s

Qual. discussion of hydrogen + hydrogen-like spectrum

Degeneracy

Modified Coulomb potential

(22) $$V = -\frac{Ze^2}{r}\left(1 + \frac{\beta}{r}\right)$$

(5) becomes $$v'' + \left[-\frac{1}{4} + \frac{A}{x} + \frac{2A\beta}{r_0}\frac{1}{x^2} - \frac{l(l+1)}{x^2}\right]v = 0$$

Put
$$l'(l'+1) = l(l+1) - \frac{2A\beta}{r_0} = l(l+1) - \frac{2\beta}{a}$$

Eq. for v becomes like (5) with $l' \to l$ (l integer, l' not integer)

Eigenvalues $A = n' + l' + 1$ (n' integer)
$$= n' + 1 + l - (l - l')$$
$$= n - (l - l') = n - \alpha_l$$

This gives

(23) $$E_{nl} = -\frac{me^4Z^2}{2\hbar^2(n-\alpha_l)^2}$$ (removes degeneracy in part)

Positive energy e.f's Radial eqn

(24) $R'' + \frac{2}{r}R' + \left\{\frac{2m}{\hbar^2}\left(E + \frac{Ze^2}{r}\right) - \frac{l(l+1)}{r^2}\right\}R = 0$

(25) $\begin{cases} R = r^l e^{ikr} F(z) & k^2 = \frac{2mE}{\hbar^2} \\ z = -2ikr \end{cases}$

Find for F

(26) $z\frac{d^2F}{dz^2} + (2l+2-z)\frac{dF}{dz} - (l+1-i\alpha)F = 0$

(27) $\alpha = me^2 Z/\hbar^2 k$

Solution is hypergeometric function

(28) $F = F(l+1-i\alpha, 2l+2, -2ikr)$

(definition & properties on next page)

Asymptotic expressions of R

(29) $\begin{cases} R_l(r\to 0) \to r^l \\ R_l(r\to\infty) \to \frac{e^{-\frac{\pi}{2}\alpha}}{(2k)^l}\frac{(2l+1)!}{|\Gamma(l+1+i\alpha)|} \times \\ \qquad \times \frac{1}{kr}\sin\left\{kr + \alpha\ln(2kr) - \frac{l\pi}{2} - \arg\Gamma\right\} \end{cases}$

For $l=0$

(30) $\begin{cases} R_0(r\to 0) \to 1 \\ R_0(r\to\infty) \to \frac{e^{-\pi\alpha/2}}{|\Gamma(1+i\alpha)|}\frac{1}{kr}\sin\left\{kr + \alpha\ln(2kr) - \arg\Gamma\right\} \end{cases}$

(31) $\begin{cases} \Gamma(n) = (n-1)! \qquad \Gamma(1+z)\Gamma(1-z) = \frac{\pi z}{\sin\pi z} \\ |\Gamma(1+i\alpha)|^2 = \frac{2\pi\alpha}{e^{\pi\alpha} - e^{-\pi\alpha}} \end{cases}$

Def. & prop. of hypergeometric fcts

(32) $$F(a,b,z) = 1 + \frac{a}{b \times 1!} z + \frac{a(a+1)}{b(b+1) \times 2!} z^2 + \ldots$$

(33) $$z F''(z) + (b-z) F'(z) - a F = 0$$

Assume b = integer z pure imaginary

Then asymptotic formula

(34) $$F(z \to i\infty) = \frac{\Gamma(b)}{\Gamma(b-a)} (-z)^{-a} + \frac{\Gamma(b)}{\Gamma(a)} z^{a-b} e^{z}$$

9- Orthogonality of wave functions.

a) One dim. case

$$(1) \quad \begin{cases} u_\ell'' + \frac{2m}{\hbar^2}\left(E_\ell - U(x)\right) u_\ell = 0 & \Big| \; u_k \\ u_k'' + \frac{2m}{\hbar^2}\left(E_k - U(x)\right) u_k = 0 & \Big| \; -u_\ell \end{cases}$$

$$u_k u_\ell'' - u_m u_k'' = \frac{d}{dx}\left(u_k u_\ell' - u_\ell u_k'\right) =$$

$$= + \frac{2m}{\hbar^2}\left(E_k - E_\ell\right) u_k u_\ell$$

$$(2) \quad \left| u_k u_\ell' - u_\ell u_k' \right|_a^b = \frac{2m}{\hbar^2}\left(E_k - E_\ell\right) \int_a^b u_k u_\ell \, dx$$

Usually $u_k, u_\ell \to 0$ for $x \to \pm\infty$.

Let then $a \to -\infty$, $b \to +\infty$

$$(3) \quad 0 = \left(E_k - E_\ell\right) \int_{-\infty}^{\infty} u_k u_\ell \, dx$$

Comments: Other types of boundary condition

e.g. Periodic

$$(4) \quad 0 = \left(E_k - E_\ell\right) \oint u_k u_\ell \, dx$$

Bounded segment (Inf. potential at $\underline{a}$ and $\underline{b}$)

$$(5) \quad 0 = \left(E_k - E_\ell\right) \int_a^b u_k u_\ell \, dx$$

In general one finds

$$(6) \quad 0 = \left(E_k - E_\ell\right) \int_{\text{domain}} u_k u_\ell \, dx$$

For

(7) $\left\{ \begin{array}{l} E_k \neq E_\ell \\ \int u_k u_\ell \, dx = 0 \end{array} \right.$

Orthogonality

In one dim. problems usually one solut. only (except for constant factor) for each eigenvalue. For normalized e.f.'s

(8) $$\int u_\ell u_k \, dx = \delta_{\ell k}$$

Developments in eigenfunctions

(9) $$\left\{ \begin{array}{l} f(x) = \sum c_k u_k(x) \\ c_k = \int_{\text{domain}} f(x) \, u_k(x) \, dx \end{array} \right.$$

b) Tridimensional case

(10) $$\left\{ \begin{array}{l} \nabla^2 u_\ell + \frac{2m}{\hbar^2}(E_\ell - U) u_\ell = 0 \\ \nabla^2 u_k + \frac{2m}{\hbar^2}(E_k - U) u_k = 0 \end{array} \right. \quad \begin{array}{l} u_k \\ -u_\ell \end{array}$$

(11) $$\nabla \cdot (u_k \nabla u_\ell - u_\ell \nabla u_k) = \frac{2m}{\hbar^2}(E_k - E_\ell) u_k u_\ell$$

(12) $$\frac{\hbar^2}{2m} \int_\sigma \left(u_k \frac{\partial u_\ell}{\partial n} - u_\ell \frac{\partial u_k}{\partial n} \right) d\sigma = (E_k - E_\ell) \int_\tau u_k u_\ell \, d\tau$$

Usually on contour of field $u_k = u_\ell \to 0$

(13) Then $(E_k - E_\ell) \int u_k u_\ell \, d\tau = 0$

or

(14) $$\int u_k u_\ell \, d\tau = 0 \quad \text{for } E_k \neq E_\ell$$

If there is one e.f. per e.v. — Normalize to 1 and then

$$\int u_k u_l \, d\tau = \delta_{kl} \quad (15)$$ (orthogonality)

Case of degeneracy. Possible to choose base such that (15) holds. (Remarks on solutions of linear diff. equations)

For example

$E_1 = E_2$ u_1 essentially $\neq u_2$

Normalize u_1 to unity

Take $u_1^{new} = u_1$

Instead of u_2 take first

$$u_2 - u_1 \int u_1 u_2 \, d\tau = u_2^{interm}$$

$u_2^{(interm)}$ is orthog. to u_1

$$\int u_1 u_2^{inter} d\tau = \int u_1 u_2 \, d\tau - \left(\int u_1^2 d\tau\right) \times \int u_1 u_2 \, d\tau$$

(where $\int u_1^2 d\tau = 1$, so the whole $= 0$)

Use then

$$u_1^{new} = u_1$$

$$u_2^{new} = \text{normalized } u_2^{interm}$$

Conclusion: even when there is degeneracy Possible & convenient to choose base such that (15) holds.

Analog of (9)

$$(16) \quad \begin{cases} f(x,y,z) = \sum c_k u_k(x,y,z) \\ c_k = \int u_k f \, d\tau \end{cases}$$

Remarks: Completeness of a set of e.f.'s — Role of complex solutions — Solution of time dependent equation

$$(17) \quad \psi = \sum c_k e^{-\frac{i}{\hbar}E_k t} u_k$$ (Meaning of $|c_k|^2$)

10 – Linear Operators

a) Functions in a field. Examples of fields (x; x, y, z; points on sph. surface; finite set of points;..)

b) Functions as vectors with infinite of finite number of dimensions

c) Operators

$$g = \mathcal{O} f \tag{1}$$

examples $g = f^2$, $g = 3f^3$, $g = \frac{df}{dx}$, $g = \frac{d^2f}{dx^2}$

$g = (7x^2+1) \times f$ etc...

Important: unit operator (indicated by 1 or I)

(2) means
$$g = 1f$$
$$g = f$$

unit operator leave function unchanged

d) In Q. M. important linear operators

Defining property (a, b constants)

$$\mathcal{O}(af+bg) = a\,\mathcal{O}f + b\,\mathcal{O}g \tag{3}$$

Examples: identity, or

$\mathcal{O} = 3$ (i.e. multiply times 3)

$\mathcal{O} = 7x^2+1$ (i.e. multiply by $7x^2+1$)

$\mathcal{O} = \frac{d}{dx}$ $\qquad$ $\mathcal{O} = \frac{d^2}{dx^2}$

Instead

$\mathcal{O}$ = take cube of

is not linear

Henceforth only lin. operators will be discussed

e) Sum and difference of operators, defined by

(4) $(A \pm B) f = Af \pm Bf$

Commutative property $A+B=B+A$

Assoc. property $A+(B+C)=(A+B)+C$ and similar are evident.

f) Product by a number

(5) $(aA)f = a(Af)$

g) Product of two operators A, B

(6) $(AB)f = A(Bf)$

Assoc. property

(7) $A(B+C) = AB+AC$ (evident)

In general however

$AB \neq BA$ (A an B do not commute)

Example $A = x$ (i.e. multiply by x)

$B = \frac{d}{dx}$

Then $(AB)f = \left(x\frac{d}{dx}\right)f = x\frac{df}{dx} = xf'$

But $(BA)f = \frac{d}{dx}(xf) = xf'+f$

h) Commutator of A and B is

(8) $AB - BA = [A, B]$

Property

(9) $[A, B] = -[B, A]$ (evident)

Example

(10) $\left[\frac{d}{dx}, x\right] = 1$ (check)

i) Powers of operator. Def. by

(11) $$A^n f = A(A \ldots A(Af))$$
$$\quad\;\; 1 \;\; 2 \quad n-1 \;\; n$$

Example $A = \frac{d}{dx}$ then $A^2 = \frac{d^2}{dx^2}$ $A^n = \frac{d^n}{dx^n}$

Property

(12) $$A^{n+m} = A^n A^m \qquad \text{(evident)}$$

(13) $$[A^n, A^m] = 0 \qquad \text{(")}$$

Two powers of same operator commute

j) Inverse operator

$$A^{-1}$$

can be defined only when

(14) $\left\{ \begin{array}{l} Af = g \\ \text{can be solved for } f \text{. Then, by definition} \\ f = A^{-1} g \end{array} \right.$

Properties

$$(A^{-1}A)f = A^{-1}(Af) = A^{-1}g = f \quad \text{that is}$$

(15) $$A^{-1}A = 1 \quad (\equiv \text{identity operator})$$

also

$$(AA^{-1})g = A(A^{-1}g) = Af = g \quad \text{that is}$$

(16) $$AA^{-1} = 1$$

And from (15) (16)

(17) $$[A, A^{-1}] = 0$$

k) Functions of an operator — Formal definition. Given a function $F(x)$ defined by analytical form (e.g. $F(x) = \sin x$, $F(x) = e^{\alpha x}$, $f(x) = \frac{x^2}{1-x}$, etc...) and operator A. Define

$$F(A) = \sum_0^\infty \frac{F^{(n)}(0)}{n!} A^n \tag{18}$$

Observe: definition not always meaningful

Example: $A = \frac{d}{dx}$,

$$e^{\alpha A} = 1 + \alpha A + \frac{\alpha^2}{2!} A^2 + \dots + \frac{\alpha^n}{n!} A^n + \dots$$

$$= 1 + \alpha \frac{d}{dx} + \dots + \frac{\alpha^n}{n!} \frac{d^n}{dx^n} + \dots = \sum_0^\infty \frac{\alpha^n}{n!} \frac{d^n}{dx^n}$$

$$e^{\alpha \frac{d}{dx}} f = \sum \frac{\alpha^n}{n!} \frac{d^n f}{dx^n} = f(x+\alpha) \tag{19}$$

Example: $A = x$ (i.e. multiply times x)

(20) Then $F(A) = F(x)$ (i.e. multiply by $F(x)$)

b) Function of two (or more) operators. Attempt to generalize (18)

$$F(A,B) = \sum_{n,m=0}^\infty \frac{F^{(n,m)}(0,0)}{n!\, m!} A^n B^m \tag{21}$$

where

$$F^{(n,m)}(x,y) = \frac{\partial^{n+m} F(x,y)}{\partial x^n \partial y^m}$$

however ambiguous except when A, B commute because otherwise e.g. $A^2B \neq ABA \neq BA^2$

Rule sometimes: symmetrize products i.e.

$$AB \to \frac{AB+BA}{2} \qquad A^2B \to \frac{A^2B + ABA + BA^2}{3} \tag{22}$$

and similar

11 – Eigenvalues and Eigenfunctions

Eigenvalue problem

$$(1) \qquad A\psi = a\psi$$

A = operator (linear)
a = number
ψ = function

ψ usually restricted to regular single-valued functions – Typical restrictions ψ(x) finite everywhere excluding infinite distance — For fields with a boundary (e.g. [illegible]) usual condition ψ vanishes on boundary

In gen. solutions only for special values of a called eigenvalues –

$$(2) \qquad A\psi_n = a_n\psi_n$$

a_n = eigenvalue
ψ_n = eigenfctn.

Example, time indep. Schrodinger eq

$$(3) \qquad \left(-\frac{\hbar^2}{2m}\nabla^2 + V\right)\psi = E\psi$$

E = eigenvalue of operator $-\frac{\hbar^2}{2m}\nabla^2 + U$
ψ = corresp. e. f.

Non degenerate e.v. when only one ψ_n except for const. factor

degenerate otherwise (double, triple, etc. degeneracy)

$a_1 \; a_2 \ldots a_n \ldots$ be all e.v.'s of (2) (each repeated times degeneracy)

$\psi_1 \; \psi_2 \ldots \psi_n \ldots$ be e.f.'s

In sect ⑨ for (3) ψ_n form orthog. system

Definition – Scalar product of f, g (functions)

(4) $(g|f) = \int g^* f$ (Observe $(g|f) = (f|g)^*$)

$\int = \int dx$ or $\int dx\,dy\,dz$ or $\sum_{\text{all points}}$

Definition g, f orthogonal when

(5) $(g|f) = 0$ or $\int g^* f = 0$

Question – When will e.f's of (2) (corresponding to different a's) be orthogonal?

Answer – When A is hermithian

Definition – Hermithian operator A (defined) has property

(6) $(g|Af) = (Ag|f)$ or

$\int g^*(Af) = \int (Ag)^* f$

Example of hermithian operators

x, $\frac{\hbar}{i}\frac{\partial}{\partial x}$, ∇^2, $-\frac{\hbar^2}{2m}\nabla^2 + U(x,y,z)$

needed appropriate boundary conditions

(7) Lemma – A hermithian → $(f|Af)$ = real number

Proof $(f|Af) = (Af|f) = (f|Af)^*$

(8) Theorem – A hermithian – E.V. real

Proof $A\psi_n = a_n \psi_n$

$(\psi_n|A\psi_n) = a_n(\psi_n|\psi_n)$ $a_n = \frac{(\psi_n|A\psi_n)}{(\psi_n|\psi_n)} = \frac{\text{real}}{\text{real}} = \text{real}$

(9) { **Theorem** A hermithian $a_n \neq a_m$ then ψ_n orthog to ψ_m

Proof

$$A\psi_n = a_n \psi_n \quad \Big| \int \psi_m^*$$

$$A\psi_m = a_m \psi_m$$

$$(A\psi_m)^* = a_m \psi_m^* \quad \Big| -\int \psi_n$$

(because a_m is real)

$$\underbrace{\int \psi_m^* A\psi_n - \int (A\psi_m)^* \psi_n}_{=0 \text{ because } A \text{ is herm.}} = (a_n - a_m) \underbrace{\int \psi_m^* \psi_n}_{(\psi_m|\psi_n)}$$

Therefore

$$(a_n - a_m)\underbrace{(\psi_m|\psi_n)}_{=0 \text{ when } a_n \neq a_m} = 0$$

QED

Quasi theorems

(10) { If $(f|Af)$ is real for all f's A is herm (inverse of (7))

(11) { If all $(\psi_n|\psi_m) = 0$ for all $a_n \neq a_m$ A is hermithian. (Inverse of (9))

These quasitheorems will be made plausible later

Normalized orthogonal e.f.s

(12) { A hermithian $a_1\ a_2 \ldots a_n \ldots$ $\psi_1\ \psi_2 \ldots \psi_n \ldots$ — ψ_r orthog to ψ_s when $a_r \neq a_s$. If there is degeneracy proceed like page 4-3

Normalization. Divide each ψ_n by $\sqrt{(\psi_n|\psi_n)}$. After **all** this for new ψ_n

(13) $$(\psi_r|\psi_s) = \delta_{rs}$$

Quasi theorem – Development of "arbitrary" f

(14) $$f = \sum c_n \psi_n \qquad c_n = (\psi_n/f)$$

or identity

(15) $$f = \sum (\psi_n|f)\psi_n$$

(Plausible later) (for all f's)

When (15) is correct (12) is called complete normalized orthogonal set.

Definition: mean value $\bar{A}$ of operator A relative to function ψ

(16) $$\bar{A} = \frac{(\psi|A\psi)}{(\psi|\psi)}$$

Example: if $A = x$ and ψ norm to 1

(17) $$\bar{x} = \int \psi^* x \psi = \int x|\psi|^2 d\tau$$

Therefore ~~wt.~~ weight used in averaging x is $|\psi|^2$

Theorem The ~~av~~ mean value of a hermitian operator is real (follows from (7) + (16))

Quasi Theorem – If the mean value of an operator relative to ~~all~~ all functions is real, the operator is hermitian (plausible later; can be proved easily from (15))

Dirac $\delta(x)$ function

(18) $\int \delta(x)\,dx = 1$ <u>when interval includes $x=0$</u>

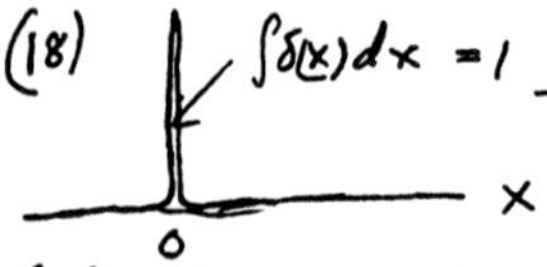

(19) $\delta(x) = \lim_{\alpha=\infty} \sqrt{\frac{\alpha}{\pi}}\, e^{-\alpha x^2}$

or

(20) $\delta x = \lim_{\alpha=\infty} \frac{\sin \alpha x}{\pi x}$

or other forms –

Properties

(21) $\int_{-\infty}^{\infty} f(x)\,\delta(x-a)\,dx = f(a)$

Take derivative respect <u>a</u>

(21) $-\int_{-\infty}^{\infty} f(x)\,\delta'(x-a) = f'(a)$

<u>Use with caution!!</u>

Fourier development

(22) $\delta(x) = \frac{1}{2\pi}\int_{-\infty}^{\infty} e^{ikx}\,dk$

Also dev. in e.f.'s (like (15))

$\delta(x-x') = \sum_n \left(\psi_n(x) \middle| \delta(x-x')\right) \psi_n(x)$ (from (21))

(23) $\delta(x-x') = \sum_n \psi_n^*(x')\,\psi_n(x)$

All six operators are hermitian

12 – Operators for mass point.

Six operators on $\psi(x,y,z)$

(1) $x\,,\ y\,,\ z\,,\ \frac{\hbar}{i}\frac{\partial}{\partial x}=p_x\,,\ \frac{\hbar}{i}\frac{\partial}{\partial y}=p_y\,,\ \frac{\hbar}{i}\frac{\partial}{\partial z}=p_z$

(a) assume ψ describes small wave packet

(n) = unit vector

$$\psi \sim e^{\frac{i}{\lambda}n\cdot x}$$

$$\lambda \approx \frac{\hbar}{mV}$$

Derive from (11–16) (fairly obvious)

(2) $\left\{\begin{array}{l}\bar{x}\,,\ \bar{y}\,,\ \bar{z} = \text{approximate coordinates of wave packet}\\ \bar{p}_x\,,\ \bar{p}_y\,,\ \bar{p}_z = \text{approximate components of mom. vector } mV\vec{n}\end{array}\right.$

(This last: $\bar{p}_x = \dfrac{\left(\psi\left|\frac{\hbar}{i}\frac{\partial\psi}{\partial x}\right.\right)}{(\psi|\psi)} \approx \dfrac{\hbar}{\lambda}n_x = mVn_x$

$\frac{\hbar}{i}\frac{\partial\psi}{\partial x}\approx\frac{\hbar}{i}\frac{i}{\lambda}n_x\psi$)

(b) (2) suggests that operators (1) have something to do with coordinates & mom. components.

Further confirmation.

Write energy (Kin + Potential of point)

(3) $E=\frac{1}{2m}\left(p_x^2+p_y^2+p_z^2\right)+U(x,y,z)=H(x,\dots,p_x\dots)$

Interpret above as function of operators (1). This operator function of operators is defined as in (10-21) but in this case definition is quite unambiguous

(4) $\begin{cases} U(x,y,z) \rightarrow \text{Operator that multiplies times } U(x,y,z) \\ p_x^2+p_y^2+p_z^2 \rightarrow \left(\frac{\hbar}{i}\right)^2\left\{\frac{\partial}{\partial x}\frac{\partial}{\partial x}+\dots\right\} \end{cases}$

$$= -\hbar^2\left(\frac{\partial^2}{\partial x^2}+\dots\right) = -\hbar^2\nabla^2$$

Therefore operator (hermitian)

(5) $H = -\frac{\hbar^2}{2m}\nabla^2 + U$

Applied to function ψ yields

(6) $H\psi = -\frac{\hbar^2}{2m}\nabla^2\psi + U\psi$

(This means merely ordinary product U times ψ)

H is called <u>energy operator</u> or <u>hamiltonian operator</u>

From previous examples, especially linear oscillator & hydrogen atom appears that

The e.v.'s of H are the energy levels of system.

Ⓒ <u>Suggested generalization</u>. <u>Postulate</u>.

Consider classical function of state of system (e.g., y coordinate; z-component of momentum; kin. energy; x component of ang. momentum & similar). All these expressible classically as functions of (x,y,z,p_x,p_y,p_z)

Form corresponding operator functions

$$\left[x;\ p_z=\frac{\hbar}{i}\frac{\partial}{\partial z};\ -\frac{\hbar^2}{2m}\nabla^2;\ M_x=yp_z-zp_y=\right.$$

$$\left.=\frac{\hbar}{i}\left(y\frac{\partial}{\partial z}-z\frac{\partial}{\partial y}\right)\ \text{and similar}\right]$$

Note: all these operators must be chosen hermithian

Postulate 1 – The only possible results of a measurement of a coordinate and momenta are the eigenvalues of the corresponding hermithian operator.

[margin note: a function $F(x\,y\,z\,p_x\,p_y\,p_z)$]

Discussion of meaning of state in classical + wave mechanics

Postulate 2 – Wave mechanical state is determined by function ψ. Its ψ variation in time according to the time dep. Sch. eq.

[margin note: However two ψ's proportional to each other represent the same state,]

Question. How does one determine the initial ψ? Answer: measure a quantity $F(\vec{x},\vec{p})$. Result of measurement will be one of the e.v.'s of F, say F_n. If F_n is non degenerate ψ immediately after the measurement is the e.f. of F corresponding to given e.v. If there is degeneracy more measurements are needed, as will be seen later.

E.v. problem

(7) $$G\, g_n(\vec{x}) = G_n\, g_n(\vec{x})$$

G = Herm. operator fct of $\vec{x}, \vec{p}$

G_n = eigenvalue (G_n is a <u>number</u>)

$g_n(x)$ = eigenfunction.

Develop ψ

(8) $$\begin{cases} \psi = \sum_n b_n\, g_n(\vec{x}) \\ b_n = (g_n|\psi) = \int g_n^* \psi \, d\tau \end{cases}$$

b_n is a number

this is state fct at time t

(9) <u>Postulate 3</u> – If $G(x,p)$ is measured probability of finding as result $G = G_n$ is proportional to $|b_n|^2$

Observe: if ψ normalized $\Sigma |b_n|^2 = 1$

Proof

$$1 = (\psi|\psi) = \left(\sum_n b_n g_n \,\middle|\, \sum_s b_s g_s\right) =$$

$$= \sum_{ns} b_n^* b_s (g_n|g_s) = \sum_{ns} b_n^* b_s \delta_{ns} = \sum_n b_n^* b_n = \Sigma |b_n|^2$$

<u>Therefore</u>: when ψ is normal. to 1

(10) $|b_n|^2$ = prob. of finding by measurement $G = G_n$

<u>Then</u>: Mean value of possible results of measuring G (ψ is normalized to 1)

$$\overline{G} = \sum_n |b_n|^2 G_n = \sum_n b_n^* G_n b_n = \sum_{sn} b_s^* G_n b_n \delta_{sn} =$$

$$= \sum_{sn} b_s^* G_n b_n (g_s | g_n) = \left(\sum_s b_s g_s \middle| \sum_n b_n G_n g_n\right) =$$

$$= \left(\psi \middle| \sum_n b_n G g_n\right) = \left(\psi \middle| G \sum_n b_n g_n\right) =$$

$$= (\psi | G\psi) = \frac{(\psi|G\psi)}{(\psi|\psi)}$$

This denominator is $= 1$

Therefore: (compare with (11-16))

Theorem. The average of op. G in the sense of (11-16) is the weighted average of of possible results that can be obtained by measuring quantity $G(\vec{x}, \vec{p})$.

Complications when e.v.'s of G are continuous

Example: op. x

$$x f(x) = x' f(x)$$

$x' =$ number

Solution $f(x) = \delta(x - x')$ = corresp. e.f.

$\delta(x - x')$ is not normalizable.

However: in sum's like (8), write $\int$ instead of $\sum$ as follows

$$n \rightarrow x'$$

$$g_n(x) \rightarrow \delta(x - x')$$

$$b_n = (g_n | \psi) \rightarrow (\delta(x - x') | \psi)\, dx'$$

$$\sum_n \rightarrow \int$$

then the inadequate normalization is compensated for by infinitesimal factor dx' and all formulas become correct

(11) Dens. of prob. of point being at $x = x'$ (8) (9)

$$\left|\left(\delta(x-x') \,|\, \psi(x)\right)\right|^2 = \left|\int \delta(x-x')\,\psi(x)\,dx\right|^2 = |\psi(x')|^2$$

← (familiar result!)

Mean value of x

(12) $$\bar{x} = (\psi | x \psi) = \int x |\psi|^2 dx$$ (ψ normalized to one)

Second example operator

(13) $$p = \frac{\hbar}{i}\frac{d}{dx}$$

e.v. equation

(14) $$p f(x) = p' f(x)$$ (p = operator, p' = number)

$$\frac{\hbar}{i} f'(x) = p' f(x)$$

general solution

(15) $$f(x) = e^{\frac{i}{\hbar} p' x}$$

(This is e.f. for eigenvalue p'. All $-\infty < p' < +\infty$ are allowable)

Again small trouble with normalization (15) not strictly normalizable — In this case sum like (8) changed as follows

(16) $$n \to p' \qquad g_n(x) \to e^{\frac{i}{\hbar} p' x} \qquad b_n = (g \,|\psi) \to (e^{\frac{i}{\hbar} p' x} | \psi)$$

$$\sum_n \to \int \frac{dp'}{2\pi\hbar}$$ (notice factor $\frac{1}{2\pi\hbar}$) this factor is needed for completness [see (11-23) and (11-22)]

$$\delta(x-x') = \sum g_n^*(x')\, g_n(x) \to \int \frac{dp'}{2\pi\hbar} e^{\frac{i}{\hbar} p'(x-x')} = \delta(x-x')$$

~~Dens of~~ Prob of finding p', $p'+dp'$

$$(18)\quad \begin{cases} \dfrac{dp'}{2\pi\hbar}\left|\left(e^{\frac{i}{\hbar}p'x}\,\middle|\,\psi(x)\right)\right|^2 & \text{(}\psi\text{ normalized)} \\ = \dfrac{dp'}{2\pi\hbar}\left|\int e^{-\frac{i}{\hbar}p'x}\,\psi(x)\,dx\right|^2 \end{cases}$$

Notice prob propost. to sq. modulus of Fourier coefficient

Check that total prob. = 1 (from (17) and normalization)

Mean value of momentum

Two expressions — From (18)

$$(19)\quad \bar{p} = \frac{1}{2\pi\hbar}\int p'\,dp'\left|\int e^{-\frac{i}{\hbar}p'x}\psi(x)\,dx\right|^2$$

or from p. 12-5 and normalization

$$(20)\quad \bar{p} = (\psi|p\psi) = \frac{\hbar}{i}(\psi|\psi') = \frac{\hbar}{i}\int\psi^*\psi'\,dx$$

part integration → $$\stackrel{\downarrow}{=} -\frac{\hbar}{i}\int\psi'^*\psi\,dx = \frac{\hbar}{2i}\int(\psi^*\psi' - \psi'^*\psi)\,dx$$

Proove: (19) & (20) are equivalent

[write (19) as double integral and use (17)]

13 – Uncertainty principle

Definite x $x = x'$ means $\psi(x) = \delta(x - x')$

Fourier has all comp with eq. amplitude

Hence no momentum limitation

(1) $$\boxed{\delta x = 0 \rightarrow \delta p = \infty}$$

Definite $p = p' \rightarrow \psi = e^{\frac{i}{\hbar} p' x}$ $|\psi|^2 = 1$

hence

(2) $$\boxed{\delta p = 0 \rightarrow \delta x = \infty}$$

Interme. case

$$\psi(x) = \begin{cases} e^{ikx} & |x| < a \\ 0 & |x| > a \end{cases}$$

(3) $$\boxed{\delta x = a}$$

−a 2a +a

From (12–18)

$$\int_{-a}^{a} e^{-\frac{i}{\hbar} p' x} e^{ikx} dx = \int_{-a}^{a} e^{i(k - \frac{p'}{\hbar})x} dx =$$

$$= \frac{\sin((p' - \hbar k)\frac{a}{\hbar})}{p' - \hbar k} \times 2\hbar$$

Prob distrib of p' is $\sim \dfrac{\sin^2 (p' - \hbar k)\frac{a}{\hbar}}{(p' - \hbar k)^2}$

$\hbar k$ $\langle \frac{2\pi\hbar}{a} \rangle$ p' therefore

(4) $\delta p' = \dfrac{\pi \hbar}{a}$

(3) + (4) ⇒

(5) $$\delta x \, \delta p \approx \hbar$$

(Uncertainty principle)

Quantitatively one proves that for any ψ

(6) $$\delta x \, \delta p \geq \frac{\hbar}{2}$$ (See Persico – Quantum Mech. p. 110 ff, p. 118)

For discussion of examples Schiff pp. 7 to 15

x + p are <u>complementary</u> according to (5)

Complementarity of time (t) and energy (E)

(7) $$\delta t \, \delta E \approx \hbar$$

has various meanings.

1) Freq. of short duration phenomenon (lasting δt) has broad band ($\delta \omega$). Find as (3) + (4)

(8) $$\delta t \, \delta \omega \approx 1$$

In wave mech. $E = \hbar \omega$, hence (7).

States of a system of short life cannot have energy more sharply defined than corresponds to (7).

2) Discussion of measurement procedures has shown that in order to measure energy accurately (δE) a time of at least $\delta t \approx \hbar/\delta E$ is needed.

All this will be discussed more sharply later

14 – Matrices

Functions in finite field (name points of field $1, 2, \ldots, n$) f is ensemble of n (complex) numbers $(f_1\ f_2 \cdots f_n)$.
Discuss: functions in continuous fields as limit of functions in a finite number of points. (e.g. describe an $f(x)$ by a table).
Consider now field of $\underline{n}$ points.

(1) $f \equiv (f_1, f_2, \ldots, f_n)$ considered as vector with complex components (n-dimensional). Limit to $n \to \infty$ (even continuous infinity) yield identification of functions with vectors in Hilbert space — Will establish theorems for finite $\underline{n}$ and in many cases results can be generalized.

Scalar product of $f \equiv (f_1 f_2 \cdots f_n)$ & $g \equiv (g_1 g_2 \cdots g_n)$

(2) $$(g|f) = \sum_1^n g_s^* f_s \quad \text{(analog of (11-4))}$$

Observe

(3) $$(g|f) = (f|g)^*$$

(4) Magnitude of "vector" $f = (f|f) = \sum_1^n |f_s|^2$

(5) Unit "vector" = "vector" of magnitude one

(6) Orthogonal vectors f & g, when $(f|g) = 0$ or equivalent $(g|f) = 0$

<u>Base</u> of <u>n</u> lin. independent "vectors"

(7) $e^{(1)}, e^{(2)}, \dots, e^{(n)}$

Condition: no linear comb. of the <u>e's</u> vanishes unless all coeff are zero. Expressed by

(8) $$\det \| e^{(i)}_k \| = 0$$

Then: any f = lin comb of <u>e's</u>

(9) $$f = \sum a_i e^{(i)}$$

(Determine coefficients a_i by solving <u>n</u> lin. eq. with det $\neq 0$)

<u>Orthonormal base</u>

when

(10) $$(e^i | e^k) = \delta_{ik}$$

If (10) then

(11) $$a_i = (e^i / f)$$

and identity

(12) $$f = \sum_i (e^i | f) e^i$$

} (evident)

<u>Operators</u>: Op. $\mathcal{O}$ is rule to convert a "vector" f into another g (in same field)

(13) $$g = \mathcal{O} f$$ (g equals $\mathcal{O}$ applied to f)

Means: components of g are functions of components of f

(14) $$g_k = \mathcal{O}_k (f_1, f_2, \dots f_n)$$

($\mathcal{O}_1, \mathcal{O}_2, \dots \mathcal{O}_n$ are n functions of n variables each defining op. $\mathcal{O}$)

<u>Linear operators</u> defined as on p. 10-1 by property

(15) $$\mathcal{O}(af + bg) = a\mathcal{O}f + b\mathcal{O}g$$

(a, b constants; f, g "vector"s)

Theorem For finite field: most general linear operator is a linear and homog. substitution

$$g = \mathcal{O} f$$

$$(16) \quad \begin{cases} g_1 = a_{11} f_1 + \dots + a_{1n} f_n \quad \text{or} \\ \dots \\ g_n = a_{n1} f_1 + \dots + a_{nn} f_n \\ g_k = \sum_{l=1}^{n} a_{kl} f_l \end{cases}$$

(a's constants)

Proof: evident that (16) is a linear operator. Prove (16) only type of linear operator. Assume $\mathcal{O}$ defined by (14) is linear. Apply (15) with

$$(17) \quad \mathcal{O}(p + \varepsilon f) = \mathcal{O}p + \varepsilon \mathcal{O} f$$

(p, f are functions, ε is infinitesimal constant)

$$(\mathcal{O}p)_k = \mathcal{O}_k(p_1, \dots, p_n)$$

$$(\mathcal{O}f)_k = \mathcal{O}_k(f_1 \dots f_n)$$

$$(\mathcal{O}(p + \varepsilon f))_k = \mathcal{O}_k(p_1 + \varepsilon f_1, \dots) =$$

$$= \mathcal{O}_k(p_1, \dots) + \varepsilon \left\{ \frac{\partial \mathcal{O}_k(p)}{\partial p_1} f_1 + \frac{\partial \mathcal{O}_k(p)}{\partial p_2} f_2 + \dots \right\}$$

Find from (17)

$$(\mathcal{O}f)_k = \sum \frac{\partial \mathcal{O}_k(p)}{\partial p_i} f_i$$

Coefficients indep. of f's, hence constants Q.E.D.

Henceforth consider only linear operators like (16)

Operator (linear) (16) represented by $n \times n$ square matrix of coefficients

$$(18) \quad O = \begin{Vmatrix} a_{11} & a_{12} & \cdots & a_{1n} \\ a_{21} & a_{22} & \cdots & a_{2n} \\ - & - & - & - \\ a_{n1} & a_{n2} & \cdots & a_{nn} \end{Vmatrix}$$

do not confuse with a determinant which is one number

Also rectangular matrices (n rows $\times$ m columns)

(e.g.) "vector" f represented by "vert. slot" matrix ($1 \times n$)

$$(19) \quad f = \begin{Vmatrix} f_1 \\ f_2 \\ \vdots \\ f_n \end{Vmatrix}$$

Algebra of matrices – Def. of operations

(20) Multiply times a number $\underline{a}$ = multiply all elements by $\underline{a}$

(21) Add & subtract (possible only for matrices that have all the same number of rows and all the same number of columns) = Matrix sum (or difference) is a matrix in which each element is the sum (or the difference) of the corresp. elements of the original matrices: example

$$\begin{vmatrix} a_{11} & a_{12} & a_{13} \\ a_{21} & a_{22} & a_{23} \end{vmatrix} + \begin{vmatrix} b_{11} & b_{12} & b_{13} \\ b_{21} & b_{22} & b_{23} \end{vmatrix} = \begin{vmatrix} a_{11}+b_{11} & a_{12}+b_{12} & a_{13}+b_{13} \\ a_{21}+b_{21} & a_{22}+b_{22} & a_{23}+b_{23} \end{vmatrix}$$

Theorems: elementary properties hold for above operations

Product of two matrices, A and B

$$(22) \quad AB = C$$

defined only when A has as many columns as B has rows. Definition

(23)
$$A = \| a_{ik} \| \quad \begin{matrix} i = 1, 2, \dots n \\ k = 1, 2, \dots m \end{matrix} \Big\} \begin{matrix} n = \text{number of rows} \\ m = \text{number of col.} \end{matrix}$$

$$B = \| b_{jl} \| \quad \begin{matrix} j = 1, 2, \dots m \\ l = 1, 2, \dots, p \end{matrix} \Big\} \begin{matrix} m = \text{no. of rows} \\ p = \text{no. of colms} \end{matrix}$$

Product $C = AB$

$$C = \| c_{rs} \| \quad \begin{matrix} r = 1, 2, \dots n \\ s = 1, 2, \dots p \end{matrix}$$

Product has as many rows as A and as many col'ms as B

(24)
$$\underset{n \times m}{A} \times \underset{m \times p}{B} = \underset{n \times p}{C}$$

Elements of product matrix obtained from rule

(25)
$$c_{rs} = \sum_{k=1}^{m} a_{rk} b_{ks}$$

(Rule of product rows × columns)

Most important special case. Product of square matrices (of equal side n) (like (18)) Then (a) product AB also is a sq. matrix of order n

(b) Product in inverted order BA can be defined and it too is sq. matrix but in general <u>different</u> from AB

$$(26)\quad \begin{cases} (AB)_{rs} = \sum_k a_{rk}\, b_{ks} \\ (BA)_{rs} = \sum_k b_{rk}\, a_{ks} \end{cases}$$

Theorem: $\det(AB) = \det(A) \times \det(B)$ evident because product of sq. matrices by same rules as rows × col. prod. of determinants

(27) Definition of commutator (for sq. matrices)

$$(28)\quad [A, B] = AB - BA$$

property: (evident) $[A, B] = -[B, A]$

Unit matrix (definition)

$$(29)\quad 1 = \begin{vmatrix} 1 & 0 & \dots & 0 \\ 0 & 1 & \dots & 0 \\ - & - & - & - \\ 0 & 0 & & 1 \end{vmatrix}$$

diagonal square matrix with all elements on main diagonal = 1

Property

$$(30)\quad \begin{cases} 1A = A1 = A \\ [1, A] = 0 \end{cases}$$

direct from (25) or (26)

Inverse of a matrix $B = A^{-1}$

Defined by

$$(31)\quad A^{-1}A = AA^{-1} = 1$$

<u>Question</u> when does inverse matrix exist?

<u>answer</u>: when $\det(A) \neq 0$ because then verify <u>rule</u>

$$(32)\quad \left(A^{-1}\right)_{rs} = \frac{\text{algebraic minor index }(s,r)\text{ in } A}{\text{determinant of } A}$$

(33) <u>Property</u> $\det\left(A^{-1}\right) = \dfrac{1}{\det(A)}$

(34) Property $[A^{-1}, A] = 0$

all this for square matrices

Property: For operator matrices like (16) all definitions of algebraic operations above are derivable from and consistent with the definitions of operator algebra given in Sect. 10. (Check one by one).

In particular define for square matrices a matrix that is a function of another matrix by same procedure of p. 10-4

Product of a square matrix by a vertical slot matrix (like (18) + (19))

(35) $\mathcal{O} f = g$ □ × ‖ = ‖

g is a vert. slot are given according to the matrix product rule (25) by equations (16).

(36) Therefore: (35) can be read with identical results either: Square matrix $\mathcal{O}$ × (vert slot f) = vert slot g

or Operator $\mathcal{O}$ applied to function f = function g

Transposed matrix of A – definition

(37) $\left\{ \begin{array}{l} A^{trans} = \text{matrix A in which rows and columns have been interchanged} \\ \text{or (equivalent)} \\ (A^{trans})_{ik} = A_{ki} \end{array} \right.$

Particular cases:

A = sq. matrix (e.g. operator matrix)
A^{trans} is obtained by changing each element with the one symmetric with respect to main diagonal

f = vert. slot (function or "vector")
f^{trans} = horizontal slot = $\| f_1, f_2, \cdots, f_n \|$

Conjugate matrix of A – definition

(38) $\left\{ \begin{array}{l} A^* = \text{matrix A in which each element is changed into its compl. conjugate} \\ (A^*)_{ik} = a^*_{ik} \end{array} \right.$

Adjoint matrix of A – (very important)

Notation for this matrix will be $\tilde{A}$

Definition

(39) $\left\{ \begin{array}{l} \tilde{A} \text{ obtained from A by transposition and conjugation} \\ (\tilde{A})_{ik} = A^*_{ki} \end{array} \right.$

Example

$$A = \begin{Vmatrix} 1 & 2+i & 3 \\ 2 & 1+i & 1-i \\ 0 & 0 & 1 \end{Vmatrix} \qquad \tilde{A} = \begin{Vmatrix} 1 & 2 & 0 \\ 2-i & 1-i & 0 \\ 3 & 1+i & 1 \end{Vmatrix}$$

Other example

(40) $f = \begin{vmatrix} f_1 \\ f_2 \\ f_3 \end{vmatrix}$ $\tilde{f} = | f_1^* \ f_2^* \ f_3^* |$

f & g are "vertical slots" i.e. functions.

$\tilde{g} f$ is then a matrix of one row and one column (See (23) & (24)) that is a number

Find

(41) $$\tilde{g} f = \sum_1^n g_s^* f_s = (g|f)$$

A, B, C, ..., K, L are matrices with such numbers of rows and columns that product matrix

(42) $P = ABC \ldots KL$ can be defined

Needed: No. of rows of each matrix = no. of columns of successive matrix

Then $\tilde{P} = \tilde{L}\tilde{K} \ldots \tilde{C}\tilde{B}\tilde{A}$

That is. The adjoint of a matrix product is the product of the adjoint matrices taken in opposite order. Proof evident from definitions.

For matrix $\tilde{g} f$ of one row and one col. of (41). adjoint is = for this case to complex conjugate

(43) $$\widetilde{\tilde{g} f} = (\tilde{g} f)^* = \tilde{f} g = (f|g)$$

15 - Hermithian matrices - Eigenvalue problems.

(1) A square ((n×n)) matrix is Hermithian when each of its elements is compl. conjugate of the one symmetric to it with respect to main diagonal. If A is hermithian

$$a_{ik} = a^*_{ki}$$

(2) Therefore a hermithian matrix is equal to its adjoint and vice versa (selfadjoint)

$$\tilde{A} = A \quad \text{when } A \text{ is hermithian}$$

All matrices

$$\begin{vmatrix} 1 & 0 \\ 0 & -1 \end{vmatrix} \quad \begin{vmatrix} 0 & 1 & 1 \\ 1 & 0 & 0 \\ 1 & 0 & 0 \end{vmatrix} \quad \begin{vmatrix} 0 & -i & e^{i\alpha} \\ i & 0 & e^{-i\beta} \\ e^{-i\alpha} & e^{i\beta} & 3 \end{vmatrix} \quad \begin{vmatrix} 0 & -i \\ i & 0 \end{vmatrix}$$

are hermithian.

(3) Observe: the diagonal elements of a hermithian matrix are real numbers

(4) Theorem (Evident from definitions). If A, B, C, ... are herm. matrices and a, b, c, ... are real numbers, then

$$aA + bB + cC + .. \quad \text{is hermithian}$$

(5) Theorem - If A is hermithian all its powers are hermithian. That is

$$A^s = \widetilde{A^s}$$

Proof: $\widetilde{A^s} = \widetilde{AA\ldots A} = \tilde{A}\tilde{A}\ldots\tilde{A} = (\tilde{A})^s = A^s$

(6) Theorem - If A is hermithian its determinant is real.

$$\det(A) = \text{real number}$$

Proof: $\det(A) = \det(A^{trans}) = [\det(\tilde{A})]^* = [\det(A)]^*$

(7) **Theorem** - If A is hermitian, so is A^{-1}

Proof: $1 = AA^{-1} = \widetilde{A^{-1}}\tilde{A} = \widetilde{A^{-1}}A$ → therefore

(because 1 is hermitian; because A is herm.)

$\widetilde{A^{-1}} = A^{-1}$

because its product with A is = 1

From these theorems follows an

(8) **Important theorem**. If F(x) is a real function of the real variable x such that for it one can define a matrix F(A) with is a function of a matrix A according to p. 14-7 and p. 10-4. Then

if A is hermitian F(A) is hermitian

because the series expansion of F(x) has real coefficients and (5)(4).

(9) If A, B are herm in general their product AB is not hermitian but symmetrized product

$$\frac{1}{2}(AB + BA) \text{ is hermitian}$$

Proof $\widetilde{\frac{1}{2}(AB+BA)} = \frac{1}{2}(\tilde{B}\tilde{A} + \tilde{A}\tilde{B}) = \frac{1}{2}(BA+AB) = \frac{1}{2}(AB+BA)$

(10) This permits in many cases to define a matrix that is a function F(A,B) of two (or more) matrices in such a way that.

If F is the symbol of a real function of its variables and A, B are hermithian,

F(A, B) is hermithian

No difficulty when A, B commute because

(11) Theorem Ⓔ A, B are herm; Ⓔ $AB = BA$ Ⓘ $P = ABAABB$ or similar products of A, B factors A or B is hermitian. (Proof: Take adjoint of P, then reorder factors using assumptions to prove $\tilde{P} = P$)

(12) Property – Def. of hermitian operators (11-(6)) is consistent with def (1) of herm. matrix. Because Ⓔ $A = \tilde{A}$ Ⓘ

$$(g \mid Af) = \tilde{g} A f = \tilde{g} \tilde{A} f = \widetilde{Ag} f = (Ag \mid f)$$

Eigenvalue problems for hermithian matrix operator

(13) Ⓔ $A = \tilde{A}$ Problem $A\psi = a\psi$ — a = eigenvalue

$$a_{11}\psi_1 + a_{12}\psi_2 + \dots + a_{1n}\psi_n = a\psi_1$$
$$a_{21}\psi_1 + a_{22}\psi_2 + \dots + a_{2n}\psi_n = a\psi_2$$
$$\text{— — —}$$
$$a_{n1}\psi_1 + a_{n2}\psi_2 + \dots + a_{nn}\psi_n = a\psi_n$$

Solvable when

$$(14)\quad \begin{vmatrix} a_{11}-a & a_{12} \dots & a_{1n} \\ a_{21} & a_{22}-a \dots & a_{2n} \\ \text{—} & \text{—} & \text{—} \\ a_{n1} & a_{n2} \dots & a_{nn}-a \end{vmatrix} = 0$$

this is determinant (not matrix)

This is algebraic equation of nth degree (Secular equation). It has n roots, some of them, however may coincide in case of degeneracy

All roots are real (Prove like (11-8))

(15) Therefore. A hermithian matrix operator has n real eigenvalues; some of them may coincide

(eigenvalues $a_1\ a_2 \ldots a_n$; functions $\psi^{(1)}\ \psi^{(2)} \ldots \psi^{(n)}$)

(16) Theorem. Eigenf. corresponding to different e.v's are orthogonal (Proof like (11-9)).

(17) Theorem. If the n roots of sec. eq. are all single then for each eigenvalue a_s there is only one ψ_s except for constant factor. (Proof given in algebra of determinants)

(18) Rule for constructing ψ_s. Substitute a_s for a in secular determinant (14). Then: The n algebraic minors of any one row of determinant are proportional to the components of vector $\psi^{(s)}$

Problem: construct the eigenvectors of

$$A = \begin{vmatrix} 0 & 1 & 0 \\ 1 & 0 & 1 \\ 0 & 1 & 0 \end{vmatrix}$$

and normalize them to 1

Same for $\begin{vmatrix} 0 & 1 \\ 1 & 0 \end{vmatrix}$ $\begin{vmatrix} 0 & -i \\ i & 0 \end{vmatrix}$ $\begin{vmatrix} 1 & 0 \\ 0 & -1 \end{vmatrix}$

(19) Case of degeneracy. An e.v. that is a solution of sec. eq. multiple of order q has q linearly independent e.f.'s — (This follows from algebra of determinants) — They can be chosen orthogonal and normalized to one.

Discuss geometrical analogy to ellipsoid

(20) Choose orthonormal set
$$\psi^{(1)}\ \psi^{(2)} \dots \psi^{(n)}; \quad \widetilde{\psi^{(r)}}\psi^{(s)} = \delta_{rs}$$
as <u>basis</u> for vector space.

(21) Development
$$f = \sum_s \left(\psi^{(s)} | f\right) \psi^{(s)}$$

This "prooves" quasitheorem (11-p.4) also proove easily all other quasi theorems of sect 11, reducing them to simple algebraic properties.

Analog of formula (11B-23). Put in (21)

$f_\rho = \delta_{\rho\sigma}$ (σ = fixed index, ρ = variable index). Then $f = \begin{vmatrix} 0 \\ 0 \\ 1 \\ 0 \\ 0 \end{vmatrix}$ (σ)

$\left(\psi^{(s)} | f\right) = \psi_\sigma^{(s)*}$. Therefore

(22)
$$\delta_{\rho\sigma} = \sum_s \psi_\sigma^{(s)*} \psi_\rho^{(s)}$$

Alternate writing of above

(23)
$$\sum_s \psi^{(s)} \widetilde{\psi^{(s)}} = 1 \quad \text{(identity } n \times n \text{ matrix)}$$

<u>Observe</u>: a matrix operator is defined by giving its eigenvectors and the corresponding eigenvalues. (Because, then)

(24)
$$Af = \sum_s a_s \left(\psi^{(s)} | f\right) \psi^{(s)} \text{ is completely defined}$$

16 – Unitary matrices – Transformations

Ⓔ A hermitian, B hermitian

(1) $\begin{pmatrix} \psi^{(1)} \cdots \psi^{(n)} \\ a_1 \cdots a_n \end{pmatrix}$ are e.f.'s and e.v.'s of A
orthonormal set

(2) $\begin{pmatrix} \varphi^{(1)} \cdots \varphi^{(n)} \\ b_1 \cdots b_n \end{pmatrix}$ for B
also orthonormal

Problem: find matrix T (transformation) that converts $\varphi^{(s)}$ into $\psi^{(s)}$

(3) $$T\varphi^{(s)} = \psi^{(s)}$$

Solution

$$T\varphi^{s}\widetilde{\varphi^{s}} = \psi^{s}\widetilde{\varphi^{s}}$$

Sum over s and use (14-23)

(4) $$T = \sum_s \psi^{s}\widetilde{\varphi^{s}}$$

Analogy with transformation of coordinates

Definition. Unitary matrix Q has defining property

(5) $$\widetilde{Q}Q = 1 \quad \text{or} \quad (\widetilde{Q} = Q^{-1})$$

(6) Theorem. T is unitary: Proof:

$$\widetilde{T} = \widetilde{\sum \psi^{s}\widetilde{\varphi^{s}}} = \sum \varphi^{s}\widetilde{\psi^{s}} \quad \text{Then using (15-20) and (15-23)}$$

$$\widetilde{T}T = \sum_{s\sigma} \varphi^{s}\widetilde{\psi^{s}}\psi^{\sigma}\widetilde{\varphi^{\sigma}} = \sum_{s\sigma} \varphi^{s}\delta_{s\sigma}\widetilde{\varphi_{\sigma}} = \sum_s \varphi^{s}\widetilde{\varphi^{s}} = 1$$

(7) **Theorem** Ⓔ T unitary

Ⓘ $(Tf|Tg) = (f|g)$

Proof: $(Tf|Tg) = \widetilde{Tf}\,Tg = \tilde{f}\,\tilde{T}\,T\,g = \tilde{f}g = (f|g)$

(8) **Theorem** Ⓔ T unitary Ⓔ $\psi^{(s)}$ an orthonormal set of n vectors

Ⓘ $T\psi^{(s)} = \varphi^{(s)}$ also form an orthonormal set.

(evident from (7))

Therefore: The unitary transformations transform an orthonormal base into another

(9) Orthonormal set $e^{(1)} = \begin{Vmatrix} 1 \\ 0 \\ \vdots \\ 0 \end{Vmatrix}$ $e^{(2)} = \begin{Vmatrix} 0 \\ 1 \\ \vdots \\ 0 \end{Vmatrix}$ $e^{(n)} = \begin{Vmatrix} 0 \\ 0 \\ \vdots \\ 1 \end{Vmatrix}$

Transformation

$$Te^{(s)} = \psi^{(s)} \quad \text{by unitary matrix}$$

$$T = \sum_s \psi^{(s)} \widetilde{e^{(s)}} = \begin{Vmatrix} \psi_1^{(1)} & \psi_1^{(2)} & \cdots & \psi_1^{(n)} \\ \psi_2^{(1)} & \psi_2^{(2)} & \cdots & \psi_2^{(n)} \\ \psi_n^{(1)} & \psi_n^{(2)} & \cdots & \psi_n^{(n)} \end{Vmatrix} \text{ or } T_{ik} = \psi_i^{(k)}$$

Transformations of coordinates of "vector" f

(10) $f = \begin{vmatrix} x_1 \\ x_2 \\ \vdots \\ x_n \end{vmatrix} = \sum x_i e^{(i)}$ to new "axes" $\psi^{(k)}$

$f = \sum x'_k \psi^{(k)}$

x_i "old" coord. of x
x'_k "new" " " x

Relationship between new and old coord.

(use (9))

(11)
$$x'_k = \widetilde{\psi}^{*} f = \sum_s \psi_s^{(k)*} x_s = (\widetilde{T})_{ks} x_s$$

or in matrix notation for vertical slots

$$x = \begin{vmatrix} x_1 \\ x_2 \\ \vdots \\ \vdots \end{vmatrix} \qquad x' = \begin{vmatrix} x'_1 \\ x'_2 \\ \vdots \\ \vdots \end{vmatrix} \qquad \begin{aligned} x' &= \widetilde{T} x = T^{-1} x \\ x &= T x' \end{aligned}$$

<u>Observe</u>: Transformation of the coordinates is the <u>inverse</u> of the transformation of the base vectors

Transformation of a matrix operator A

<u>Question</u>. The matrix operator A defines a linear substit. on the coord. x of a vector. What is the corresponding linear subst. A' on the coordinates x' of same vector?

<u>Answer</u>: from (11) $x = T x'$; from definition of question above

$$\underbrace{A x}_{= A T x'} = T A' x'$$

$\rightarrow$ $T^{-1} A T x' = A' x'$ for an arbitrary x'. Therefore

(12)
$$\boxed{A' = T^{-1} A T = \widetilde{T} A T}$$

or inverse

$$A = T A' T^{-1} = T A' \widetilde{T}$$

$\underline{A}$ is transformed into $\underline{A}'$ by T

Properties (E)

$$A' = T^{-1}AT$$
$$B' = T^{-1}BT$$
$$\text{- - - -}$$

(13) Then

$$A' \pm B' = T^{-1}(A \pm B)T$$
$$A'B' = T^{-1}(AB)T$$
$$A'^{n} = T^{-1}A^{n}T$$
$$F(A') = T^{-1}F(A)T$$

also $1 = T^{-1}1T$

and similar properties. Verify directly

The algebra of $A', B', \ldots$ is identical to the algebra of $A, B, \ldots$

(14) Also: A' has the same e.v's of $\underline{A}$. and its e.f.'s are

$$\psi'^{(s)} = T^{-1}\psi^{(s)} = \tilde{T}\psi^{(s)}$$ (check)

$$\text{or } T\psi'^{(s)} = \psi^{(s)}$$

(15) Trace or Spur of a matrix A (sq. matrix)

$$Sp(A) = \sum_{1}^{n} A_{ss}$$ (sum of elements of main diagonal)

(16) Theorem A & A' have same spur

$$Sp\,A' = Sp\,\tilde{T}AT = \sum_{ikr} (\tilde{T})_{ik} A_{kr} T_{ri} =$$
$$= \sum_{kr} A_{kr} (T\tilde{T})_{rk} = \sum A_{kr}\delta_{kr} = \sum A_{kk} = Sp\,A$$

~~Problem.~~

Problem.

(17) A hermitian, T unitary $A' = \tilde{T} A T$

Determine T such that A' is diagonal.

Answer

$$T = \sum_s \psi^{(s)} \widetilde{e^{(s)}} \quad \text{(see (9))}$$

Because

$$A' = \tilde{T} A T = \sum_{s\sigma} e^s \widetilde{\psi^s} \underbrace{A \psi^\sigma}_{= a_\sigma \psi^\sigma} \widetilde{e^\sigma} = \sum_{s\sigma} a_\sigma e^s \underbrace{\widetilde{\psi^s} \psi^\sigma}_{= \delta_{s\sigma}} \widetilde{e^\sigma}$$

$$= \sum_s a_s e^s \widetilde{e^s} = \sum_s a_s \begin{Vmatrix} 0 & 0 & 0 & 0 \\ 0 & 0 & 0 & 0 \\ 0 & 0 & 1 & 0 \\ 0 & 0 & 0 & 0 \\ 0 & 0 & 0 & 0 \end{Vmatrix} (s) = \begin{Vmatrix} a_1 & 0 & & 0 \\ 0 & a_2 \ldots & & 0 \\ & & \ddots & \\ 0 & 0 & & a_n \end{Vmatrix}$$

A is transformed in a diagonal matrix A' with the e.v.'s on main diagonal.

T transforms the original base $e^{(s)}$ into $\psi^{(s)}$

This means: A is made diagonal by taking its e.f.'s as the ~~base~~ new ~~coordinates~~ coordinates base

(18) Theorem

$$\text{Spur}(A) = \sum_1^n a_s$$

Evident from previous and (16)

(19) New definition of a matrix F(A). Three steps:

one Convert A to diagonal A' as in (17)

$$A' = \tilde{T} A T$$
$$A = T A' \tilde{T}$$

two

$$F(A') = \begin{vmatrix} F(a_1) & 0 & 0 \ldots \\ 0 & F(a_2) & 0 \ldots \\ 0 & 0 & F(a_3) \ldots \end{vmatrix}$$

three

$$F(A) = T\, F(A')\, \tilde{T}$$

Proove easily (using (13)) — Definition (19) is equivalent to gen. definition of p. 10-4 whenever that definition is meaningful. But Definition (19) does not restrict F.

(20) Theorem

$$[A, F(A)] = 0$$

even when def. (19) is used. Proof: $[A', F(A')] = 0$ because both diagonal, then use (13)

(21) Theorem (Inverse of (20)

If A, B commute and A is non degenerate

$$B = F(A)$$

Proof: Transform A into diag. matrix A' as in (17)

$$A' = \tilde{T} A T = \begin{vmatrix} a_1 & 0 & \\ 0 & a_2 & \\ 0 & 0 & \ldots \end{vmatrix}$$

$$B' = \tilde{T} B T$$

From $[A, B] = 0$ follows $[A' B'] = 0$

$$[A', B']_{ik} = (a_i - a_k)\, b'_{ik} = 0$$

From this and $a_i \neq a_k$ for $i \neq k$ follows $b'_{ik} = 0$ for $i \neq k$

Therefore B' also diagonal $= \begin{vmatrix} b_1 & 0 & 0 \ldots \\ 0 & b_2 & 0 \ldots \\ 0 & 0 & b_3 \ldots \end{vmatrix} = B'$

Therefore $B' = F(A')$ provided F is one of the infinite fcs for which $F(a_1) = b_1,\ F(a_2) = b_2 \ldots F(a_n) = b_n$

Transform back + use (19) to proove (21).
Incidentally we have proved:

(22) **Theorem**: A diagonal, non degenerate B, commutes with A. Then: also B must be diagonal

(23) If A in (22) is degenerate then B does not have to be diagonal. But B has the structure shown in the following example easily generalized

$$A = \begin{pmatrix} a_1 & 0 & 0 & 0 & 0 \\ 0 & a_1 & 0 & 0 & 0 \\ 0 & 0 & a_2 & 0 & 0 \\ 0 & 0 & 0 & a_2 & 0 \\ 0 & 0 & 0 & 0 & a_2 \end{pmatrix} \qquad B = \left(\begin{array}{cc|ccc} b_{11} & b_{12} & & 0 & \\ b_{21} & b_{22} & & & \\ \hline & & b_{33} & b_{34} & b_{35} \\ & 0 & b_{43} & b_{44} & b_{45} \\ & & b_{53} & b_{54} & b_{55} \end{array}\right)$$

(24) This has <u>important application</u>.
Assume: A, B hermithian and $[A,B]=0$
Solve the e.v. probleme of A as on p. 15-3.
Then transform A into a diagonal matrix
$A' = \tilde{T} A T$ as in (17). Also $B' = \tilde{T} B T$. A' and B' commute. Then:
If A is non degenerate, by (22) B' is diagonal and the e.v. problem of B is solved
If A is degenerate, then B' is of form like in example (23) and its secular equation splits into simpler equations each having order = to the degree of degeneracy of the e.v's of A.

17 – Observables

Observable = function of state of system.

1 – In q.m. one constructs for each observable Q a linear operator (also Q). If the observable is essentially real, Q is a hermithian operator

2 – A measurement of Q may yield as value of Q only one of the e.v.'s of op. Q

(1) $Qf_{q'} = q'f_{q'}$ (q' is e.v. $f_{q'}$ is e.fctn)

3 – State of system represented by ψ (usually normalized to 1) factor immaterial

4 – How to determine ψ?

Measure Q, find $Q = q'$

Then if q' non degenerate,

(2) $\psi = f_{q'}$

If q' <u>is degenerate</u> then

ψ = linear comb. of all e.f's corresponding to q'

~~Then~~ (Vector ψ belongs to subspace q')

(3) $Q\psi = q'\psi$ defines the subspace q'

In order to determine ψ within subspace q' choose observable P that commute with Q

(4) $$[P,Q]=0$$

(5) Theorem: Ⓔ $[P,Q]=0$; Ⓔ $Q\psi = q'\psi$, i.e. ψ belongs to subspace q' ; Ⓓ $P\psi$ also belongs to subspace q', i.e. $Q(P\psi)=q'(P\psi)$.
Proof: $Q(P\psi) = QP\psi = PQ\psi = Pq'\psi = q'(P\psi)$

Consider P as operator within subspace q'. I will have e.v's & e.f.'s in number equal to the dimension of subspace q' obtained as simultaneous solutions of

(6) $$\begin{cases} Q\psi = q'\psi \\ P\psi = p'\psi \end{cases}$$ p' = e.v. of P within subspace $Q=q'$

(6) defines a sub-sub-space $(Q=q', P=p')$. If this is onedimensional (6) defines ψ except for factor. Otherwise ψ is limited to sub-sub-space. Then measure also another observable R such that

(7) $$[R,Q]=0 \quad [R,P]=0$$

R operates in sub sub space

(8) $$Q\psi = q'\psi \quad P\psi = p'\psi \quad R\psi = r'\psi$$

Define sub-sub-sub-space. If it has <u>one</u> dimension ψ is determined. If not, go on.

5 – If ψ is known and A is measured:
Prob of finding $A=a'$ is $|(\varphi_{a'}|\psi)|^2$

6– Time variation of "state vector" ψ

H = hamiltonian operator (Hermitian). Then time dependent Schroedinger eq.

$$(9) \qquad i\hbar\dot{\psi} = H\psi$$

Observe

$$(10) \qquad -i\hbar\dot{\tilde{\psi}} = \tilde{\psi}\tilde{H} = \tilde{\psi}H$$

(11) <u>Theorem</u>: $\tilde{\psi}\psi$ (i.e the normalization constant) is a time constant. Therefore: if $\psi(0)$ is normalized, so is $\psi(t)$.

Proof:

$$\frac{d}{dt}\tilde{\psi}\psi = \tilde{\psi}\dot{\psi} + \dot{\tilde{\psi}}\psi \overset{(9)\,\&\,(10)}{=} \frac{1}{i\hbar}\tilde{\psi}H\psi - \frac{1}{i\hbar}\tilde{\psi}H\psi = 0$$

(12) 7– If classically

$$H = H(q_1, q_2, \cdots, p_1, p_2, \cdots)$$

$H_{operator}$ substituting $p_1 = \frac{\hbar}{i}\frac{\partial}{\partial q_1}, \cdots$

<u>but</u> not always unambiguous

These operators on functions $f(q_1\, q_2 \cdots q_n)$

Very infinite "index" $q_1', q_2', \cdots, q_s'$

8– Transformation to matrix.

Frequently convenient to transform to orthonormal base using the e.f's of some pertinent operator like hamiltonian or unpert. hamiltonian. Assume *one* q only ($q=x$)

Orthonormal base functions

(13) $$\psi^{(1)}(x),\ \psi^{(2)}(x),\ \dots,\ \psi^{(n)}(x),\ \dots$$

Transf. unitary matrix (See p. 16-2)

(14) $$T = \left\| \begin{matrix} \psi^{(1)}(x') & \psi^{(2)}(x') \dots & \psi^{(n)}(x') \dots \\ \psi^{(1)}(x'') & \psi^{(2)}(x'') \dots & \psi^{(n)}(x'') \dots \\ \psi^{(1)}(x''') & \psi^{(2)}(x''') \dots & \psi^{(n)}(x''') \dots \\ - - - & - - & - - - \end{matrix} \right\|$$

Doubly infinite matrix!!

horizontal index $1, 2, \dots n, \dots$ (may or may not be discrete)

vert. index x', x'', x''' (all values of x, usually continuous infinity)

(<u>Handle with caution!</u>)

a "vector or function" $f(x) = \sum \varphi_n^* \psi^{(n)}$

$$\varphi_n = (\psi^{(n)} | f) = \int \psi^{(n)*} f\, dx = \widetilde{\psi}^n f$$

(15) $\begin{cases} f(x')\ f(x'')\ f(x''') & \text{old coordinates of } f \\ \varphi_1\ \ \varphi_2\ \ \varphi_n & \text{new coordinates of } f \end{cases}$

(16) Operator A transforms to $\widetilde{T} A T$

$$A = \left| \begin{matrix} A_{11} & A_{12} & A_{1m} \dots \\ A_{21} & A_{22} & A_{2m} \dots \\ A_{31} & A_{32} & A_{3m} \dots \\ - - & - & - \end{matrix} \right| \qquad A_{nm} = (\psi^{(n)} | A \psi^{(m)}) = \int \psi^{(n)*}(x) A \psi^{(m)}(x)\, dx$$

If A is hermitian $A_{nm} = A_{mn}^*$

(17)
A_{nm} = matrix element of A between states n & m. Also

$$A_{nm} = \langle \psi^{(n)} | A | \psi^{(m)} \rangle = \langle n | A | m \rangle$$

$$\psi^{(m)} = |m\rangle = \text{Ket} \quad \tilde{\psi}^{n} = \langle n| = \text{brac}$$

Example – Take

(18)
$\psi^{(n)}(x) = u_n(x)$ = e.f's of oscillator (4–17)

They are e.f.s of operator

$$H = \frac{1}{2m}p^2 + \frac{m\omega^2}{2}x^2$$

After unitary transf. (14) H transforms to diag. matrix

(19)

$$H = \begin{vmatrix} \frac{\hbar\omega}{2} & 0 & 0 & 0 & \cdots \\ 0 & \frac{3}{2}\hbar\omega & 0 & 0 & \cdots \\ 0 & 0 & \frac{5}{2}\hbar\omega & 0 & \cdots \\ 0 & 0 & 0 & \frac{7}{2}\hbar\omega & \cdots \\ \cdots & & & & \end{vmatrix}$$

$$H_{nm} = H_{nn}\delta_{nm} = \hbar\omega\left(n + \frac{1}{2}\right)\delta_{nm}$$

Determine matrix $\underline{x}$ and matrix $\underline{p}$.

(20)
From (18) & $px - xp = \hbar/i$

$$\frac{\hbar}{im}p = Hx - xH \quad \underline{\text{or}} \quad \frac{\hbar}{im}p_{rs} = (Hx - xH)_{rs} = (H_{rr} - H_{ss})x_{rs} = \hbar\omega(r-s)x_{rs}$$

(21) From $Hp - pH = -\frac{\hbar}{i} m\omega^2 x$

$$\therefore \frac{\hbar}{i} m\omega^2 x_{rs} = \hbar\omega(r-s)\, p_{rs}$$

Combine to find

$$x_{rs} = (r-s)^2 x_{rs}$$

Therefore $x_{rs} \neq 0$ only for $r = s \pm 1$

(22) Also $p_{rs} \neq 0$ " " "

Also

$$p_{r,r+1} = -i m\omega\, x_{r,r+1}$$

Determine $|x_{r,r+1}|^2 + |x_{r-1,r}|^2 = \frac{\hbar\omega}{m\omega^2}(r+\tfrac{1}{2})$ from (18) (19) (22). Find

$$|x_{r,r+1}|^2 = \frac{\hbar}{2m\omega}(r+1)$$

Discuss arbitrariness of argument

(23)
$$x_{r,r+1} = x_{r+1,r} = \sqrt{\frac{\hbar}{2m\omega}}\sqrt{r+1}$$

$$p_{r,r+1} = -p_{r+1,r} = -i\sqrt{\frac{\hbar m\omega}{2}}\sqrt{r+1}$$

(24)
$$x = \sqrt{\frac{\hbar}{2m\omega}}\begin{vmatrix} 0 & \sqrt{1} & 0 & 0 & \cdots \\ \sqrt{1} & 0 & \sqrt{2} & 0 & \cdots \\ 0 & \sqrt{2} & 0 & \sqrt{3} & \cdots \\ 0 & 0 & \sqrt{3} & 0 & \cdots \\ \cdot & \cdot & \cdot & \cdot & \cdot \end{vmatrix} ; \quad p = \sqrt{\frac{\hbar m\omega}{2}}\begin{vmatrix} 0 & -i\sqrt{1} & 0 & 0 & \cdots \\ i\sqrt{1} & 0 & -i\sqrt{2} & 0 & \cdots \\ 0 & i\sqrt{2} & 0 & -i\sqrt{3} & \cdots \\ 0 & 0 & i\sqrt{3} & 0 & \cdots \\ \cdot & \cdot & \cdot & & \end{vmatrix}$$

Check $px - xp = \frac{\hbar}{i}$

Important linear combinations

$$(25)\quad \begin{cases} \tilde{a} = \sqrt{\dfrac{m\omega}{2\hbar}}\,x - \dfrac{i}{\sqrt{2\hbar m\omega}}\,p = \begin{vmatrix} 0 & 0 & 0 & 0 & \cdots \\ \sqrt{1} & 0 & 0 & 0 & \cdots \\ 0 & \sqrt{2} & 0 & 0 & \cdots \\ 0 & 0 & \sqrt{3} & 0 & \cdots \\ \cdots & & & & \end{vmatrix} \\ a = \sqrt{\dfrac{m\omega}{2\hbar}}\,x + \dfrac{i}{\sqrt{2\hbar m\omega}}\,p = \begin{vmatrix} 0 & \sqrt{1} & 0 & 0 & - \\ 0 & 0 & \sqrt{2} & 0 & - \\ 0 & 0 & 0 & \sqrt{3} & - \\ 0 & 0 & 0 & 0 & \sqrt{4} \cdots \end{vmatrix} \end{cases}$$

a, $\tilde{a}$ are <u>non</u> hermitian operators (destruction & creation operators of field theory).

Check commutation relation

$$(26)\qquad a\tilde{a} - \tilde{a}a = 1$$

18 – The angular momentum

(1) $\vec{M} = \vec{x} \times \vec{p}$

(2)
$$\begin{cases} M_x = yp_z - zp_y = X \\ M_y = zp_x - xp_z = Y \\ M_z = xp_y - yp_x = Z \end{cases}$$

(3) $M^2 = M_x^2 + M_y^2 + M_z^2$

Prove easily commutation rules

(4)
$$\begin{cases} [M_x, M_y] = \frac{i\hbar}{\mathcal{O}} M_z \,; \quad [M_y, M_z] = \frac{i\hbar}{\mathcal{O}} M_x \\ [M_z, M_x] = \frac{i\hbar}{\mathcal{O}} M_y \end{cases}$$

or

(5) $\vec{M} \times \vec{M} = \frac{i\hbar}{\mathcal{O}} \vec{M}$

(6) $[M_x, M^2] = [M_y, M^2] = [M_z, M^2] = 0$

(7) $[r^2, M_x] = [r^2, M_y] = [r^2 M_z] = 0$

(8) $[r^2, M^2] = 0$

Use units $\hbar = 1$

(9) $[X, Y] = +iZ \quad [Y, Z] = +iX \quad [Z, X] = +iY$

Take representation with M^2 diagonal matrix

Find e.v. of M^2. From (2) & (3) expressed in polar coordinates

$$(10)\quad \begin{cases} M_z = \frac{\hbar}{i}\frac{\partial}{\partial\varphi} \\ M^2 = -\hbar^2 \Lambda \end{cases}$$

Therefore.

$$(11)\quad \begin{cases} M^2 \text{ has e.v.'s } \hbar^2 l(l+1) & l = 0,1,2\ldots \\ M_z \quad \text{"} \quad \text{"} \quad \hbar m & m = \ldots,-2,-1,0,1,2,\ldots \end{cases}$$

$$(12)\quad \begin{cases} \text{e.f.'s of } M^2 \quad \boxed{\hbar = 1} \\ M^2 = l(l+1) \qquad \psi = f(r)\, Y_{lm}(\vartheta,\varphi) \\ 2l+1 \text{ – fold degeneracy, in addition to } r\text{-degeneracy} \end{cases}$$

$$(13)\quad \begin{cases} \text{For each } M^2 = l(l+1) \text{ find} \\ M_z = m = (l, l-1, l-2, \ldots, -l) \end{cases}$$

Partial matrices M_x, M_y, M_z

$$(14)\quad M_z = \hbar \begin{vmatrix} l & 0 & 0 & \cdots \\ 0 & l-1 & 0 & \cdots \\ 0 & 0 & l-2 & \cdots \\ \cdot & \cdot & \cdot & \cdot \\ 0 & 0 & 0 & -l \end{vmatrix} ; \quad M_x = \frac{\hbar}{2} \begin{vmatrix} 0 & b_l & 0 & 0 & - & 0 & 0 \\ b_l & 0 & b_{l-1} & 0 & - & 0 & 0 \\ 0 & b_{l-1} & 0 & b_{l-2} & - & 0 & 0 \\ 0 & 0 & b_{l-2} & 0 & - & 0 & 0 \\ - & - & - & - & - & - & - \\ 0 & 0 & 0 & 0 & - & 0 & b_{-l+1} \\ 0 & 0 & 0 & 0 & - & b_{-l+1} & 0 \end{vmatrix}$$

$$M_y = \frac{\hbar}{2} \begin{vmatrix} 0 & -ib_l & 0 & 0 & \cdots & 0 & 0 \\ ib_l & 0 & -ib_{l-1} & 0 & \cdots & 0 & 0 \\ 0 & ib_{l-1} & 0 & -ib_{l-1} & \cdots & 0 & 0 \\ - & - & - & - & - & - & - \\ 0 & 0 & 0 & 0 & & 0 & -ib_{-l+1} \\ 0 & 0 & 0 & 0 & & ib_{-l+1} & 0 \end{vmatrix}$$

$$b_s = \sqrt{(l+s)(l+1-s)}$$

(See Schiff: p. 144)

Proove directly either from properties of spherical harmonics! — Or from commutation rules. Further more general discussion of ang. momentum later.

$$(15)\ \left\{ l=0 \qquad M^2=0 \qquad M_z = M_x = M_y = \|0\| \right.$$

$$(16)\ \begin{cases} l=1 \qquad M^2=2 \qquad M_z=\begin{vmatrix} 1 & 0 & 0 \\ 0 & 0 & 0 \\ 0 & 0 & -1 \end{vmatrix} \quad M_x=\begin{vmatrix} 0 & \frac{1}{\sqrt{2}} & 0 \\ \frac{1}{\sqrt{2}} & 0 & \frac{1}{\sqrt{2}} \\ 0 & \frac{1}{\sqrt{2}} & 0 \end{vmatrix} \\ M_x+iM_y=\begin{vmatrix} 0 & \sqrt{2} & 0 \\ 0 & 0 & \sqrt{2} \\ 0 & 0 & 0 \end{vmatrix} \qquad M_y=\begin{vmatrix} 0 & -i/\sqrt{2} & 0 \\ i/\sqrt{2} & 0 & -i/\sqrt{2} \\ 0 & i/\sqrt{2} & 0 \end{vmatrix} \\ M_x-iM_y=\begin{vmatrix} 0 & 0 & 0 \\ \sqrt{2} & 0 & 0 \\ 0 & \sqrt{2} & 0 \end{vmatrix} \end{cases}$$

Non hermithian lin. combinations

$$(17)\ \begin{cases} \frac{1}{\hbar}\langle m+1 | M_x+iM_y | m\rangle = \sqrt{(l+m+1)(l-m)} \\ \frac{1}{\hbar}\langle m-1 | M_x-iM_y | m\rangle = \sqrt{(l+m)(l+1-m)} \end{cases}$$

all other matrix elements vanish!

$$(18)\ \begin{cases} \text{Observe: operator } M_x+iM_y \text{ changes} \\ \text{state } |m\rangle \longrightarrow \sqrt{(l+m+1)(l-m)}\ |m+1\rangle \\ (M_x-iM_y)\ |m\rangle \longrightarrow \sqrt{(l+m)(l+1-m)}\ |m-1\rangle \end{cases}$$

M_x+iM_y increases, M_x-iM_y decreases the m value by one unit.

19 – Time dependence of observables – Heisenberg representation.

Time dependent equation

$$(1) \qquad i\hbar\dot{\psi} = H\psi$$

May be used to define following unitary transformation (function of time)

$$(2) \qquad S(t)$$

$S(t)$ transforms a vector $\varphi(0)$, referred to $t=0$ into a vector $\varphi(t)$, referred to time t. $\varphi(t)$ is obtained by integrating

$$(3) \qquad i\hbar\dot{\varphi} = H\varphi$$

between 0 and t taking $\varphi(0)$ as initial value of φ.

Already prooved (17 – p. 3) that $S(t)$ is unitary.

$$(4) \qquad \begin{cases} \varphi(t) = S(t)\,\varphi(0) \\ \varphi(0) = S(t)^{-1}\,\varphi(t) = \widetilde{S(t)}\,\varphi(t) \end{cases}$$

In particular for wave function

$$(5) \qquad \begin{cases} \psi(t) = S(t)\,\psi(0) \\ \psi(0) = \widetilde{S(t)}\,\psi(t) \end{cases}$$

When H is time independent, explicit expression of $S(t)$

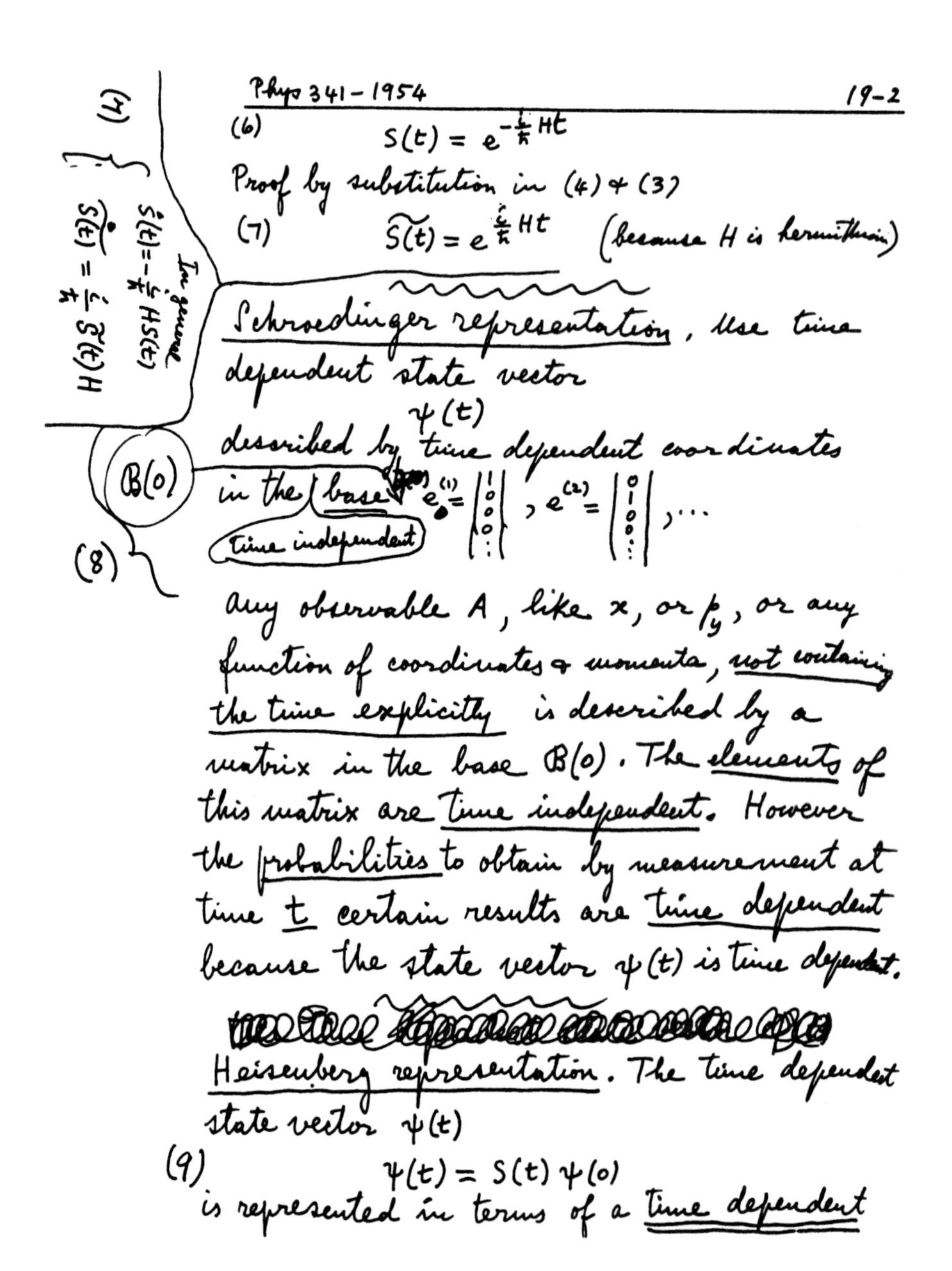

(6) $$S(t) = e^{-\frac{i}{\hbar}Ht}$$

Proof by substitution in (4) & (3)

(7) $$\tilde{S}(t) = e^{\frac{i}{\hbar}Ht}$$ (because H is hermitian)

(4) In general $\dot{S}(t) = -\frac{i}{\hbar} H S(t)$

$\dot{\tilde{S}}(t) = \frac{i}{\hbar}\tilde{S}(t) H$

Schroedinger representation, use time dependent state vector

$\psi(t)$

described by time dependent coordinates in the base $\mathcal{B}(0)$ (time independent) $e^{(1)} = \begin{vmatrix} 1 \\ 0 \\ 0 \\ 0 \\ \vdots \end{vmatrix}$, $e^{(2)} = \begin{vmatrix} 0 \\ 1 \\ 0 \\ 0 \\ \vdots \end{vmatrix}$, ...

(8)

any observable A, like x, or p_y, or any function of coordinates & momenta, not containing the time explicitly is described by a matrix in the base $\mathcal{B}(0)$. The elements of this matrix are time independent. However the probabilities to obtain by measurement at time t certain results are time dependent because the state vector $\psi(t)$ is time dependent.

Heisenberg representation. The time dependent state vector $\psi(t)$

(9) $$\psi(t) = S(t)\,\psi(0)$$

is represented in terms of a time dependent

(10) set of base vectors

$$e^{(s)}(t) = S(t)\,e^{(s)}$$

(Base $\mathcal{B}(t)$)

(11) The coordinates of $\psi(t)$ in $\mathcal{B}(t)$ are time independent and equal to the coordinates of $\psi(0)$ in $\mathcal{B}(0)$. Because:

$$\widetilde{e^{s}(t)}\,\psi(t) = \widetilde{S(t)\,e^{(s)}}\,S(t)\,\psi(0) = \widetilde{e^{(s)}}\,\tilde{S}\,S\,\psi(0) = \widetilde{e^{(s)}}\,\psi(0)$$

This is sometimes abbreviated in the statement that the state vector is time independent. Rather the state vector is referred to a set of coordinates that follows it in its motion and it appears constant when referred to such coordinates.

The matrix elements of observable A function of coordinates & momenta but not containing $\underline{t}$ explicitly are time constants in the base $\mathcal{B}(0)$ but not in the Heisenberg time dependent base $\mathcal{B}(t)$.

The matrix representing A becomes

$$A(t) = \tilde{S}(t)\,A\,S(t)\,;\quad A = S\,A(t)\,\tilde{S} \tag{12}$$

where A is the time independent matrix representing the observable in the Schroedinger base $\mathcal{B}(0)$.

Find

$$\frac{d}{dt}A(t) = \tilde{S}(t)\,A\,\dot{S}(t) + \dot{\tilde{S}}(t)\,A\,S(t) \overset{\text{use (7)}}{=} \frac{i}{\hbar}(\tilde{S}HAS - \tilde{S}AHS)$$

Put like (12)

$$H(t) = \tilde{S} H S \tag{13}$$

Find then

$$\frac{dA(t)}{dt} = \frac{i}{\hbar}\left[H(t), A(t)\right] \tag{14}$$

This is the <u>Heisenberg equation of motion</u> for operators that do not explicitly depend on time.

Meaning of $A(t)$: measuring operator $A(t)$ on state $\psi(0)$ at $t=0$ is equivalent to measure A on future state $\psi(t)$

→ If H does not contain t explicitly, from (14) find

$$\frac{dH(t)}{dt} = \frac{i}{\hbar}\left[H(t), H(t)\right] = 0 \quad \text{i.e.}$$

$$H(t) = \text{constant} = H(0) = H \tag{15}$$

This however is correct only <u>provided</u> the hamiltonian does not contain the time explicitly.

Relationship between (14) & the Hamilton eq.'s

Assume

$$H = H(q_1 q_2 \ldots p_1 p_2 p_3 \ldots) \quad \text{(time independent)}$$

(16) $[p_s, q_s] = \frac{\hbar}{i}$ leads (in simple cases) to $[H, q_s] = \frac{\hbar}{i}\frac{\partial H}{\partial p_s}$

$[H, p_s] = -\frac{\hbar}{i}\frac{\partial H}{\partial q_s}$. Then from (14)

$$\frac{dq_s}{dt} = \frac{i}{\hbar}[H, q_s] = \frac{\partial H}{\partial p_s}\,; \quad \frac{dp_s}{dt} = \frac{i}{\hbar}[H, p_s] = -\frac{\partial H}{\partial q_s}$$

– Hamilton equations

20 – Conservation theorems.

(1) { Assume in this section H does not contain t explicitly

(2) { same assumption for other operators A, B, C ...

Then: According to (19 – (15))

(3) { H is constant (conservation of energy

(4) { Similarly from (19 – (14)), A is conserved when

$$[H, A] = 0$$

Meaning: measuring A now or at a future time gives same result.

Classical conservation theorems of momentum and ang. momentum are related to symmetry properties of physical space. i.e.

Conserv. of momentum ⟷ Translation symmetry

" " angular momentum ⟷ Rotation symmetry

Assume symmetry ~~property~~ operations of system.

Examples: Translation (case of internal forces only)

Rotation (case of internal forces only or of central forces for rotation around ~~center~~ source of central forces)

Rotation around z-axis (whenever it applies)
Reflection on a plane of symmetry.

For each such case introduce operator T

(5) Defined
$$Tf(\text{positions}) = f(\text{positions changed by symmetry operation})$$

Example: operation = reflection about xy plane
$$Tf(x_1, y_1, z_1, x_2, y_2, z_2, \ldots) = f(x_1, y_1, -z_1, x_2, y_2, -z_2, \ldots)$$

(6) Theorem: T is unitary: evident because T obviously conserves the normalization of f
$$\tilde{T}T = 1$$

(7) Theorem: T commutes with H
$$[H, T] = 0$$

Because consider one e.v. E_n of H defining a vector sub-space of the (one or more) e.f's of H belonging to E_n – T operates within the subspace – This means: the matrix elements T_{rs} of T in the H representation vanish for $E_r \neq E_s$. Which is equivalent to (7)

(8) Theorem
$$[H, \tilde{T}] = 0$$
Because $\tilde{T} = T^{-1}$ is also a symmetry operation (inverse of T)

Theorem. A unitary matrix T has e.f's that are orthogonal (like those of a hermithian matrix), and e.v.'s of modulus 1.

Proof:

$$T = \frac{T+\tilde{T}}{2} + i\,\frac{T-\tilde{T}}{2i}$$

(these are hermithian and commute)

therefore they have a common set of e.f's that are orthogonal. They are also the e.f's of T. (First part of theorem). Take these eigenvectors as base and reduce T to diagonal form. Then from $T\tilde{T}=1$ follows that diagonal elements have modulus 1 (Second part of theorem).

(9) Therefore: (α_s is real)

e.v's of	T	$e^{i\alpha_s}$
e.v's of	$\tilde{T}$	$e^{-i\alpha_s}$
e.v's of	$\frac{T+\tilde{T}}{2}$	$\cos\alpha_s$
e.v's of	$\frac{T-\tilde{T}}{2i}$	$\sin\alpha_s$

all belong to same wave fct $\psi^{(s)}$

(10) All above 4 matrices commute with each other and with H. Therefore they are time constants and and their wave functions $\psi^{(s)}$ may be chosen to coincide with the eigenfunctions of the energy

<u>Symmetry Group</u> is the ensemble of all the transformations corresponding to a certain symmetry property: E.g. all the rotations of the x, y, z-axes form the rotations group

Comments on group theory & Q. M.

(11) <u>Representation of a group</u> = ensemble of unitary matrices corresponding to all operations of group and having same algebra.

(12) <u>Irreducible representation</u> = representation that cannot be transformed to $\begin{pmatrix} \blacksquare & 0 \\ 0 & \blacksquare \end{pmatrix}$ for all its matrices at same time.

(13) <u>Property</u>: Irred. repres. are determined uniquely by the abstract structure of the group

(14) <u>Usually useful to</u>. Choose a set of base vectors $\varphi^{(1)}\ \varphi^{(2)} \dots$ that split into sub sets $\varphi^{(\ell 1)}\ \varphi^{(\ell 2)} \dots \varphi^{(\ell g)}$ each one of which (set ℓ) is transformed into itself by all operations of the symmetry group according to one of its irreducible representations R_ℓ

(15) <u>Wigner theorem</u>. If a quantity $\underline{A}$ commutes with all operations of a group (e.g. the Hamiltonian), the matrix elements of A $\widetilde{\varphi^{(i)}} A \varphi^{(k)}$ for the above choice of base vectors vanish when the two vectors $\varphi^{(i)}, \varphi^{(k)}$ correspond to different irred. repres. Otherwise

$$\langle \varphi^{(\ell i)} | A | \varphi^{(\lambda k)} \rangle = a_{\ell,\lambda}\, \delta_{ik}$$ with $a_{\ell\lambda}$ a number provided $R_\ell = R_\lambda$

Application 1 – Translation symmetry and the conservation of momentum. For systems with internal forces only – (Means homogeneity of physical space)

(16) $T(\vec{a}) = T(a, b, c)$ = translations by $[(a, b, c) \equiv \vec{a}]$

Observe: all these T's corresponding to $\vec{a}$ $\vec{a'}$ commute among themselves and of course with H. (Abelian group). Therefore: choose representation in which H + all T's are orthogonal. For a wave function ψ then

(17)
$$T(\vec{a})\psi = e^{i\alpha(\vec{a})}\psi$$
$\alpha(\vec{a})$ is a function of the vector $\vec{a}$

From
$$T(\vec{a})T(\vec{a'}) = T(\vec{a} + \vec{a}')$$
conclude
$$\alpha(\vec{a}) + \alpha(\vec{a'}) = \alpha(\vec{a} + \vec{a'})$$
i.e.
$$\alpha = \vec{k} \cdot \vec{a} = k_x a + k_y b + k_z c$$

$\vec{k}$ is a constant vector for the given wave function ψ. It would be different for another wave function.

(18) $T(\vec{a}) = e^{i\vec{k}\cdot\vec{a}}$ is an irreducible representation of the translation group

(19) Find: $\hbar k$ = momentum of system. Proof:
Take an infinitesimal translation by ε along x
$(a = \varepsilon, b = 0, c = 0)$ $\quad T = e^{ik_x\varepsilon} = 1 + ik_x\varepsilon$

$$T\psi(x_1 y_1 z_1 x_2 y_2 z_2 \ldots) = (1 + ik_x\varepsilon)\psi = \psi + ik_x\varepsilon\psi$$
$$\psi(x_1+\varepsilon, y_1, z_1, x_2+\varepsilon, y_2, z_2, \ldots) = \psi + \varepsilon\left(\frac{\partial\psi}{\partial x_1} + \frac{\partial\psi}{\partial x_2} + \ldots\right)$$

(20)
$$k_x\psi = \frac{1}{i}\left(\frac{\partial\psi}{\partial x_1} + \frac{\partial\psi}{\partial x_2} + \ldots\right) = \frac{1}{\hbar}\left(p_x^{(1)}\psi + p_x^{(2)}\psi + \ldots\right)$$
$$\hbar k_x = \sum_s p_x^{(s)} \qquad \hbar k = \sum_s p^{(s)}$$
s summed to all mass points

Wave functions of $\vec{p}$

(21)
$$\psi = e^{\frac{i}{\hbar}\vec{p}\cdot\vec{x}_1}\,\varphi(\vec{x}_2-\vec{x}_1,\ \vec{x}_3-\vec{x}_1,\ \dots)$$

p here is a vector with components that are <u>numbers, not operators</u>. They are the e.v.'s of the operators $p_x,\ p_y,\ p_z$.

(22) In it $\vec{p}=0$ and φ is a function of the relative coordinates <u>only</u>. Comments on greater generality.

Frequently one makes a transformation to a moving system of reference (Galilei an or Lorentz as case may be) in order to reduce system to c. of m. frame (barsy).

<u>Application 2</u> – <u>Rotation symmetry</u> & the conserv. of <u>angular momentum</u>

For systems with internal forces only or also with external central forces. Center of rotation in this case is the origin of the central forces.

Take

(23)
T = rot. by infinitesimal ω_z around z axis

$$x \to x-\omega_z y \qquad y \to y+\omega_z x\ , \quad z \to z$$

$$T\psi(x_1\ y_1\ z_1\ x_2\ y_2\ z_2\dots) = \psi(x_1-\omega_z y_1,\ y_1+\omega_z x_1,\ z_1,\dots$$

Form hermithian operator

$$M_z = \frac{\hbar}{\omega_z}\,\frac{T-\tilde{T}}{2i}$$

Also similarly M_x and M_y and

(24)
$$M^2 = M_x^2 + M_y^2 + M_z^2$$

Follows:

$$(25)\left\{ \begin{array}{l} M_x\ ,\ M_y\ ,\ M_z\ ,\ M^2 \\ \text{Are constants of motion. (Conservation of ang. mom.)} \end{array}\right.$$

Also from their definition follow the commutation relations

$$(26)\quad \left\{ \begin{array}{l} [M_x, M_y] = \frac{\hbar}{i} M_z \quad \text{\& similar or} \\ \vec{M} \times \vec{M} = \frac{\hbar}{i} \vec{M} \\ [M_x, M^2] = 0 \quad \text{\& similar} \end{array}\right.$$

like for the ang. mom of a single point (p 18-1)

One proves that the matrix structure of (15) found in (18–(12)(13)(14)(17)(18)) follows from commutation rules only and obtains therefore for (15) <u>with the following exception</u>. In sect. 18 l was an integral number. In general, however, also half odd values of l are allowable. This is important for the quantum theory of spin.

The transformation $T(\alpha)$ corresponding to a rotation by α around z is

$$(27)\quad T(\alpha)\psi = e^{im\alpha}\psi$$

in the representation in which M_z and M^2 are diagonal, m = integral or half integral

<u>Application 3</u> = <u>Reflection symmetry & conservation of parity</u>. For systems with internal & central forces only one postulates reflection symmetry

T corresponds to $x \to -x \quad y \to -y \quad z \to -z$ reflection about the origin. This implies that right & left are physically equivalent.

(28) $T\psi(x_1, y_1, z_1, x_2, y_2, z_2, \ldots) = \psi(-x_1, -y_1, -z_1, -x_2, -y_2, -z_2, \ldots)$

Observe

(29) $$T^2 = 1$$

Also T commutes with the operators (25) and of course with H.

(30) { Normally choose eigenfunctions of M_1^2, M_z, and T

(they all intercommute). Because of (29) the e.v.'s of T, which in general are given by (9) become:

(31) e.v.'s of T are ± 1

This permits classification of states

(32) { even for $T = +1$; odd for $T = -1$ } (parity)

The parity is a property that does not change <u>as long as only central & internal forces</u> act on system.

21 – Time independent perturbation theory.

(1) $$H = \underbrace{H_0}_{\text{unpert.}} + \underbrace{\mathcal{H}}_{\text{perturbation}}$$

(2) $$H_0 u_0^{(n)} = E_0^{(n)} u_0^{(n)}$$

(3) $$H = H_0 + \lambda \mathcal{H} \qquad \lambda \to 1 \text{ at end}$$

(4) $$u^{(n)} = u_0^{(n)} + \lambda u_1^{(n)} + \lambda^2 u_2^{(n)} + \cdots$$

(5) $$E^{(n)} = E_0^{(n)} + \lambda E_1^{(n)} + \lambda^2 E_2^{(n)} + \cdots$$

(6) $$(H_0 + \lambda \mathcal{H}) u^{(n)} = E^{(n)} u^{(n)}$$

(7) $$H_0 u_0^{(n)} = E_0^{(n)} u_0^{(n)} \quad \leftarrow \text{This is (2)}$$

(8) $$H_0 u_1^{(n)} - E_0^{(n)} u_1^{(n)} - E_1^{(n)} u_0^{(n)} = -\mathcal{H} u_0^{(n)}$$

(9) $$H_0 u_2^{(n)} - E_0^{(n)} u_2^{(n)} - E_2^{(n)} u_0^{(n)} = -\mathcal{H} u_1^{(n)} + E_1^{(n)} u_1^{(n)}$$

- - - - - (comment on this)

(10) Put
$$u_1^{(n)} = \sum{}' c_{nm}^{(1)} u_0^{(m)}$$
$$u_2^{(n)} = \sum{}' c_{nm}^{(2)} u_0^{(m)}$$

Substitute in (8), (9) using (2) or (7)

(11) $$\sum_m{}' c_{nm}^{(1)} \left(E_0^{(m)} - E_0^{(n)}\right) u_0^{(m)} - E_0^{(n)} u_0^{(n)} = -\mathcal{H} u_0^{(n)}$$

(12) $$\sum_m{}' c_{nm}^{(2)} \left(E_m^{(0)} - E_n^{(0)}\right) u_0^{(m)} - E_1^{(n)} u_0^{(n)} = -\mathcal{H} u_1^{(n)} + E_1^{(n)} u_1^{(n)}$$

Matrix element

$$(13)\quad \mathcal{H}_{mn} = \left(u_0^{(m)} \middle| \mathcal{H}\, u_0^{(n)}\right) = \langle m | \mathcal{H} | n \rangle = \int u_0^{m*} \mathcal{H} u_0^{n}\, dx = \widetilde{u_0^{(m)}}\, \mathcal{H}\, u_0^{(n)}$$

ⓐ Determine $E_1^{(n)}$. Multiply (11) by $\widetilde{u_0^{(n)}}$ to left, use orthogonality

$$(14)\quad \widetilde{u_0^{n}}\, u^{m} = \delta_{nm}$$

$$(15)\quad E_1^{(n)} = \widetilde{u_0^{(n)}}\, \mathcal{H}\, u_0^{(n)} = \mathcal{H}_{nn}$$

~~Pertu~~ First order perturbation of of energy is mean value of $\mathcal{H}$ over unperturbed state.

Next $\widetilde{u_0^{(m)}} \times (11)$ yields

$$(16)\quad c_{nm}^{(1)} = \frac{\mathcal{H}_{mn}}{E_0^{(n)} - E_0^{(m)}}$$

or e.f's to first order

$$(17)\quad u_0^{(n)} + \sum_m{}' \frac{\mathcal{H}_{mn}}{E_0^{(n)} - E_0^{(m)}}\, u_0^{(m)}$$

Same treatment on (12) yields

$$(18)\quad E_2^{(n)} = \sum_m{}' \frac{\mathcal{H}_{nm}\mathcal{H}_{mn}}{E_0^{(n)} - E_0^{(m)}} = \sum_m{}' \frac{|\mathcal{H}_{nm}|^2}{E_0^{(n)} - E_0^{(m)}}$$

$$(19)\quad c_{nm}^{(2)} = \sum_s{}' \frac{\mathcal{H}_{ms}\mathcal{H}_{sn}}{(E_0^{n} - E_0^{s})(E_0^{n} - E_0^{m})} - \frac{\mathcal{H}_{mn}\mathcal{H}_{nn}}{(E_0^{n} - E_0^{m})^2}$$

Example – Lin. oscillator perturbed by const. force F

(20) $$\mathcal{H} = -Fx$$

(21) From (p. 17-6)

$$\begin{cases} \mathcal{H}_{nm} = -F x_{nm} \\ x_{n,n+1} = \sqrt{\frac{\hbar}{2m\omega}}\sqrt{n+1} \\ x_{n,n-1} = \sqrt{\frac{\hbar}{2m\omega}}\sqrt{n} \end{cases} \qquad E_0^{(n)} = \hbar\omega\left(n+\tfrac{1}{2}\right)$$

$$\ldots = x_{n,n-3} = x_{n,n-2} = x_{nn} = x_{n,n+2} = x_{n,n+3} = \ldots = 0$$

Therefore pert of energy. First order

(22) $$E_1^{(n)} = \mathcal{H}_{nn} = -F x_{nn} = 0$$

Second order

(23) $$\begin{cases} E_2^{(n)} = \sum' \frac{|\mathcal{H}_{nm}|^2}{E_0^n - E_0^m} = F^2\left(\frac{|x_{n,n+1}|^2}{-\hbar\omega} + \frac{|x_{n,n-1}|^2}{\hbar\omega}\right) = \\ = \frac{F^2}{\hbar\omega}\left(-\frac{\hbar}{2m\omega}(n+1) + \frac{\hbar}{2m\omega}n\right) = -\frac{F^2}{2m\omega^2} \end{cases}$$

Energy of all states is decreased by $F^2/(2m\omega^2)$

Direct proof

(24) $$H = \frac{1}{2m}p^2 + \frac{m\omega^2}{2}x^2 - Fx = \frac{1}{2m}p^2 + \frac{m\omega^2}{2}\left(x - \frac{F}{m\omega^2}\right)^2 - \frac{F^2}{2m\omega^2}$$

↑ shift of eq. position; ← correction of energy as above

Example – Zeeman effect (no spin) ($p \to p - \frac{e}{c}A$)

(25) $H = \frac{1}{2M}\left(p - \frac{e}{c}A\right)^2 + U(r)$ A = vect. pot, $H = \nabla \times A$

$= \frac{1}{2M} p^2 + U(r) - \frac{e}{2MC}\, p \cdot A$ + quadr. terms in A neglected

(comment: $p \cdot A - A \cdot p = \frac{\hbar}{i} \nabla \cdot A = 0$ in static case)

Mag. field // to z, intensity B

(26) $A_x = -\frac{B}{2} y, \quad A_y = \frac{B}{2} x, \quad A_z = 0$

(27) $H = \overbrace{\frac{1}{2M} p^2 + U(r)}^{H_0} - \overbrace{\frac{eB}{2MC}(x p_y - y p_x)}^{\mathcal{H}}$

Unpert. e.f.'s

(28) $u_{n,l,m}(r, \vartheta, \varphi) = R_{nl}(r)\, Y_{lm}(\vartheta, \varphi)$

In this case pert. theory trivial because (28) are also e.f's of (27).

(29) $$\begin{cases} H_0 u_{nlm} = E^{(0)}_{nl} u_{nlm} \\ \mathcal{H} u_{nlm} = -\frac{eB}{2MC} m\, u_{nlm} \\ E_{nlm} = E^{(0)}_{nl} - \frac{eB}{2MC} m \end{cases}$$

[Diagram: $m=0 \to$ one level; $m=1 \to$ three levels; $m=2 \to$ five levels; $m=0 \to$ one level]

Discussion (Selection rule $m \to m \pm 1$, $\to m$.

also corresp. principle)

Discuss role of constants of motion in limiting types of unpert. e.f's that enter into perturbation sums.

Bohr magneton.

Write perturbation repr. int. of orbit and field in (27)

$$(30)\quad \begin{cases} \mathcal{H} = -\vec{B}\cdot\vec{\mu} & \vec{\mu} = \text{magn. mom. of orbit} \\ \vec{\mu} = \dfrac{e\hbar}{2mc}\left(\dfrac{1}{\hbar}\vec{M}\right) & \dfrac{\vec{M}}{\hbar} = \text{ang. mom. of orbit in } \hbar \text{ units.} \end{cases}$$

(31) Interpret: to each unit $\hbar$ of ang. momentum of the orbit there is associated a unit

$$\mu_0 = \frac{e\hbar}{2mc} = 9.2732\times 10^{-21}\ \text{cm}^{5/2}\ \text{gr}^{1/2}\ \text{sec}^{-1}$$

of magnetic moment (μ_0 = Bohr magneton).

Topics for discussion.

Proof of (31) from classical orbit model

Proof of (31) from current density derived from continuity equation (2-(7)) and (2-(9)))

$$(32)\quad J = \frac{\hbar e}{2imc}\left(\psi^*\nabla\psi - \psi\nabla\psi^*\right)$$

$$(33)\quad \begin{cases} \mu_z = \int \frac{1}{2}(\vec{x}\times J)_z\, d^3x \\ \psi = F(r,\vartheta)e^{im\varphi} \qquad \psi^* = F(r,\vartheta)e^{-im\varphi} \\ \int |\psi|^2 d^3x = 1 \end{cases}$$

$$(34)\quad \longrightarrow \mu_z = \frac{e\hbar}{2mc}m$$

Ritz Method. From (22). ψ approximates exact $\psi^{(n)}$ with error of first order. Then

$$(35)\quad \overline{H} = (\psi|H\psi) = \tilde{\psi} H\psi = \int \psi^* H \psi \, dx$$

approximates $E^{(n)}$ with error of second order.

(36) Practical application: Guess wave function. Compute $\tilde{\psi} H \psi$. If guess of ψ is fair guess of E is good.

More precisely. Theorem: Minimum problem

$$(37)\quad \delta(\tilde{\psi} H \psi) = 0 \quad \text{with condition } \tilde{\psi}\psi = 1$$

Leads to Schrödinger equation

(38) Proof $\widetilde{\delta\psi} H\psi + \tilde{\psi} H \delta\psi - \lambda \tilde{\psi} \delta\psi - \lambda \widetilde{\delta\psi}\, \psi = 0$

$$\widetilde{\delta\psi}\,(H\psi - \lambda\psi) + \widetilde{(H\psi - \lambda\psi)}\, \delta\psi = 0$$

leads to equation

$$H\psi = \lambda\psi \quad (= \text{Schrod. eq. with } E = \lambda)$$

Therefore: Solve min. problem (37). The min. value is the lowest e.v., extremal values are other e.v's.

Practical application: Choose reasonable guess for $\psi^{(0)} \approx f(x, \alpha, \beta, \ldots)$. $\alpha, \beta, \ldots$ are adjustable parameters. Compute

$$(39)\quad E(\alpha, \beta, \ldots) = \frac{\int f^*(x, \alpha, \ldots)\, H f(x, \alpha, \ldots)\, dx}{\int f^*(x, \alpha, \ldots)\, f(x, \alpha, \ldots)\, dx}$$

Find values of $\alpha, \beta, \ldots$ that

(40) $$E(\alpha, \beta, \ldots) = \min$$

The min value of E is close to lowest energy level, $f(x, \alpha, \beta, \ldots)$ is fair approx. to e.f.

Example. Oscillator problem

(41) $$H = \frac{1}{2}p^2 + \frac{1}{2}x^2 \qquad \hbar = 1 \quad m = 1 \quad \omega = 1$$

Trial $f(x)$

$\langle \alpha \rangle$ 1

Find

(42) $$E(\alpha) = \frac{\frac{1}{2}\int_{-\alpha}^{\alpha} x^2 f^2(x)\, dx - \frac{1}{2}\int_{-\alpha}^{\alpha} f(x) f''(x)\, dx}{\int_{-\alpha}^{\alpha} f^2(x)\, dx} = \frac{\frac{\alpha^3}{30} + \frac{1}{\alpha}}{\frac{2}{3}\alpha} = \frac{1}{20}\alpha^2 + \frac{3}{2}\frac{1}{\alpha^2}$$

(43) Min at $\alpha = \sqrt[4]{30} = 2.34$

$E(2.34) = 0.548$, within 10% of correct lowest e.v. 0.500000

Prove: $E(\alpha, \beta, \ldots)$ given by (29) obeys

(44) $$E(\alpha, \beta, \ldots) \geq E_0$$

with E_0 = lowest en. e.v. (For proof develop f in e.f's of H)

Discussion of practical use.

22 - Case of degeneracy or quasi degeneracy

Perturbation procedure of Sect 21 breaks down when $E_0^{(n)} - E_0^{(m)} = 0$ or very small. (See 21 (18) and (21-16))

(1) unpert. e.f's
$u_0^{(1)} \; u_0^{(2)} \ldots u_0^{(g)}$ — These deg. or quasi degenerate for unp. problem
$u_0^{(g+1)} \; u_0^{(g+2)} \ldots$ — These have unpert. energies quite different from previous.

Search for solutions (of first order approx) of type

$$u = \sum_1^g c_s u_0^{(s)} + \sum_{g+1}^{\infty} c_\alpha u_0^{(\alpha)} = \qquad (2)$$

c_α small of first order
c_s large

$H = H_0 + \mathcal{H}$

$Hu = Eu \qquad E = E_0 + \varepsilon$

In first approximation

$$\sum_1^g c_s (H - E) u_0^{(s)} + \sum_{g+1} c_\alpha (H_0 - E_0) u_0^{(\alpha)} = 0 \qquad (3)$$

$$\sum_{g+1}^{\infty} c_\alpha (E_0^{(\alpha)} - E_0) u_0^{(\alpha)}$$

Multiply by $\widetilde{u_0^{(l)}}$ to left, $l = 1, 2, \ldots, g$.

(4)
$$\sum_1^g c_s (H_{ls} - E\delta_{ls}) = 0$$
This is secular problem of order g

$$\begin{vmatrix} H_{11}-E & H_{12} \cdots & H_{1g} \\ H_{21} & H_{22}-E \cdots & H_{2g} \\ - & - \; - & - \\ H_{g1} & H_{g2} \cdots & H_{gg}-E \end{vmatrix} = 0$$

Determines the g energy levels corresp. to the degenerate or quasi deg. set of g levels of unpert problem.

Determine then

$$(5)\qquad c_\alpha = \frac{\sum_1^g c_\sigma H_{\alpha\sigma}}{E_0 - E_0^{(\alpha)}}$$ <u>large denominator!</u>

gives first order correction to wave function.

<u>Comments</u>: role of conservation theorems in reducing secular problem (4)

<u>Example</u> Stark effect in H $n=2$ levels

Perturbation

$$(6)\qquad \mathcal{H} = +eFz \qquad F = \text{electric field}$$

4 deg levels of unpert. problem

$$(7)\qquad 2s,\ 2p_1,\ 2p_0,\ 2p_{-1}$$ (see p. 8-4)

Observe:

$$(8)\qquad [\mathcal{H}, M_z] = 0$$

Therefore perturbation mixes only states of equal <u>m</u>, like $2s$ and $2p_0$.

$2p_1$ and $2p_{-1}$ have their energies perturbed (as in case of non degeneracy) in first approx. (21-(15)) by amt

$$(9)\begin{cases} \langle 2p_1 | eFz | 2p_1 \rangle = \\ = eF\int z|\psi_{2p_1}|^2 d^3x = 0 \end{cases}$$ (because z odd, $|\psi_{2p_1}|^2$ even)

Same for $2p_{-1}$

Therefore $2p_1$ & $2p_{-1}$ unperturbed in first approximation.

(10) $$\psi_{2s} = \frac{1}{\sqrt{32\pi a^3}}\left(2-\frac{r}{a}\right)e^{-\frac{r}{2a}}$$

(11) $$\psi_{2p_0} = \frac{1}{\sqrt{32\pi a^3}}\frac{r}{a}e^{-\frac{r}{2a}}\cos\vartheta$$

$$\langle 2s|z|2s\rangle = \langle 2p_0|z|2p_0\rangle = 0$$

(12) $$\begin{cases} \langle 2s|z|2p_0\rangle = \dfrac{1}{32\pi a^3}\displaystyle\int_0^\infty\int_0^\pi\left(2-\frac{r}{a}\right)\frac{r}{a}e^{-\frac{r}{a}}r\cos^2\vartheta\, 2\pi r^2 dr \sin\vartheta\, d\vartheta \\ = \dfrac{1}{16a^3}\underbrace{\displaystyle\int_0^\infty\left(2-\frac{r}{a}\right)\frac{r}{a}r^3 e^{-\frac{r}{a}}dr}_{-72a^4}\underbrace{\displaystyle\int_0^\pi\cos^2\vartheta\sin\vartheta\, d\vartheta}_{2/3} = -3a \end{cases}$$

Perturb matrix

(13) $$eF\begin{vmatrix} 0 & -3a \\ -3a & 0 \end{vmatrix} \text{ has e.v.'s } \pm 3eFa$$

Therefore in

(14)

Energy level to first approx.	E.f of zero approx.
$-\frac{me^4}{2\hbar^2}\frac{1}{4}$	ψ_{2p_1}
$-\frac{me^4}{2\hbar^2}\frac{1}{4}$	$\psi_{2p_{-1}}$
$-\frac{me^4}{2\hbar^2}\frac{1}{4}+3eFa$	$\frac{1}{\sqrt{2}}(\psi_{2s}+\psi_{2p_0})$
$-\frac{me^4}{2\hbar^2}\frac{1}{4}-3eFa$	$\frac{1}{\sqrt{2}}(\psi_{2s}-\psi_{2p_0})$

23 – Time dependent perturbation theory, Born approximation.

(1) $\left\{ H = H_0 + \mathcal{H} \right.$ H_0 time independent; $\mathcal{H}$ may be time dependent

Unperturbed Sch. eq.

(2) $$i\hbar \dot{\psi}_0 = H_0 \psi_0$$

has solution

(3) $$\psi_0 = \sum a_n^{(0)} u_0^{(n)} e^{-\frac{i}{\hbar} E_0^{(n)} t}$$

(4) $a_n^{(0)}$ constants. $\quad H_0 u_0^{(n)} = E_0^{(n)} u_0^{(n)}$

Solve Schr eq

(5) $$i\hbar \dot{\psi} = (H_0 + \mathcal{H})\psi$$

by

(6) $$\psi = \sum a_n(t) u_0^{(n)} e^{-\frac{i}{\hbar} E_0^{(n)} t}$$

→ then multiply by $\widetilde{u_0^{(s)}}$ to left + use orthonormality and (4).

(7) $$\dot{a}_s = -\frac{i}{\hbar} \sum_n a_n \langle s|\mathcal{H}|n\rangle e^{\frac{i}{\hbar}(E_0^{(s)} - E_0^{(n)})t}$$

(8) $$\langle s|\mathcal{H}|n\rangle = \widetilde{u_0^{(s)}} \mathcal{H} u_0^{(n)} = \int u_0^{(s)*} \mathcal{H} u_0^{(n)} dx = \mathcal{H}_{sn}$$

(7) is exact. Use it approximately by substituting in right hand side $a_n(0)$ for $a_n(t)$. Then

(9) $$a_s(t) \approx a_s(0) - \frac{i}{\hbar} \sum_n a_n(0) \int_0^t \mathcal{H}_{sn}(t) e^{\frac{i}{\hbar}(E_0^{(s)} - E_0^{(n)})t} dt$$

<u>Important special case</u>, at $t=0$ system in state n. Then $a_n(0)=1$, all other a's are zero.

$$(10)\quad (s\neq n)\quad a_s(t) = -\frac{i}{\hbar}\int_0^t \mathcal{H}_{sn}(t)\, e^{\frac{i}{\hbar}(E_0^{(s)}-E_0^{(n)})t}\, dt$$

Matrix element $\mathcal{H}_{sn}(t)$ causes transitions $n \to s$.

<u>Transitions</u> from n <u>to a continuum of states</u>

Assume $\mathcal{H}_{sn}$ indep. of time, then

$$(11)\quad \begin{cases} a_s(t) = -\mathcal{H}_{sn}\dfrac{e^{\frac{i}{\hbar}(E_0^s-E_0^n)t}-1}{E_0^s-E_0^n} \\ |a_s(t)|^2 = 4|\mathcal{H}_{sn}|^2 \dfrac{\sin^2\frac{t}{2\hbar}(E_0^{(s)}-E_0^{(n)})}{(E_0^{(s)}-E_0^{(n)})^2} \end{cases}$$

Prob of transition <u>to one</u> state s

$$(12)\quad \begin{cases} P(t) = \sum_s |a_s(t)|^2 = 4\overline{|\mathcal{H}_{sn}|^2}\sum \dfrac{\sin^2\frac{t}{2\hbar}(E^s-E^n)}{(E^s-E^n)^2} = \\ = 4\overline{|\mathcal{H}_{sn}|^2}\rho(E_n)\underbrace{\displaystyle\int \dfrac{\sin^2\frac{t}{2\hbar}(E^s-E^n)\, d(E^s-E^n)}{(E^s-E^n)^2}}_{\frac{\pi t}{2\hbar}} \\ = t\,\dfrac{2\pi}{\hbar}\overline{|\mathcal{H}_{sn}|^2}\rho(E_n) \end{cases}$$

$$\int \frac{\sin^2\alpha x}{x^2}dx = \frac{\pi}{\alpha}$$

$$(13)\ \&\quad \begin{cases} \rho(E_n) = \text{no of states } s\text{, close to } E_n \text{ per unit energy interval.} \\ \boxed{\text{Rate of transition} = \dfrac{2\pi}{\hbar}\overline{|\mathcal{H}_{sn}|^2}\rho(E_n)} \end{cases}$$

Discuss: distribution of final states as function of t & relation with uncertainty principle

Example : Born approximation.

(14) Scattering by a potential $U(\vec{x})$

initial $\xrightarrow{p}$ $U(x)$ final p', ϑ

$|p'| = |p|$

$U(x) = \mathcal{H}$ treated as perturbation

(15) initial state $\dfrac{1}{\sqrt{\Omega}} e^{\frac{i}{\hbar}\vec{p}\cdot\vec{x}}$ (Ω = vol. of box)

final state $\dfrac{1}{\sqrt{\Omega}} e^{\frac{i}{\hbar}\vec{p}'\cdot\vec{x}}$

$$\langle p'|\mathcal{H}|p\rangle = \frac{1}{\Omega}\int U(x)\, e^{\frac{i}{\hbar}(\vec{p}-\vec{p}')\cdot\vec{x}}\, d^3x$$

$$= \frac{1}{\Omega} U_{p-p'} \quad \text{Fourier transform of } U$$

(16) No of final states in solid angle $d\omega$ per unit energy interval

$$\rho_{d\omega} = \frac{\Omega\, d\omega}{(2\pi\hbar)^3}\frac{p^2\,dp}{v\,dp} = \frac{\Omega p^2}{8\pi^3\hbar^3 v}\, d\omega$$

v = velocity $\quad v\,dp = dE$ (correct also relativistically)

Rate of transitions into $d\omega$

$$d\omega\, \frac{v}{\Omega}\frac{d\sigma}{d\omega} = \frac{2\pi}{\hbar}\left|\frac{1}{\Omega}U_{p-p'}\right|^2 \frac{\Omega p^2}{8\pi^3\hbar^3 v}\, d\omega$$

(17)
$$\boxed{\frac{d\sigma}{d\omega} = \frac{1}{4\pi^2\hbar^4}\frac{p^2}{v^2}\left|U_{p-p'}\right|^2}$$

(18) For non relativistic mechanics $m = \dfrac{p}{v}$

$$\frac{d\sigma}{d\omega} = \frac{m^2}{4\pi^2\hbar^4}\left|U_{p-p'}\right|^2$$

Limits of validity (discuss)

(19) $$\frac{1}{\hbar} L\left(\sqrt{p^2 \pm 2mU} - p\right) \ll 1$$

< L >

$\hat{U}$

Scattering by Coulomb center

$$U = \frac{zZe^2}{r}$$

(20) $$U_{p-p'} = zZe^2 \int \frac{e^{\frac{i}{\hbar}(\vec{p}-\vec{p}')\cdot\vec{x}}}{r}\, d^3x = \frac{4\pi zZe^2}{\frac{1}{\hbar^2}|\vec{p}-\vec{p}'|^2} = \frac{4\pi\hbar^2 zZe^2}{4p^2\sin^2\frac{\vartheta}{2}}$$

(use $\nabla^2\varphi = -4\pi\frac{e^{i\alpha x}}{r}$)

(21) $$\frac{d\sigma}{d\omega} = \frac{z^2Z^2}{4}\left(\frac{me^2}{p^2}\right)^2\frac{1}{\sin^4\frac{\vartheta}{2}}$$ (Identical to classical Rutherford formula)

Suggested discussion topics.

Scattering by potential well – Nuclear forces

Limit of long wave length – isotropic scattering

" " short " " – forward "

Role of the mass (neutrino)

Exponential decay of original state in case (11)

24 - Emission and absorption of radiation.

(1) $$\mathcal{H} = eBz\cos\omega t$$

B = amplitude.

At $t=0$ atom in state n. From (23-(10))

(2) $$a_m(t) = -\frac{i}{\hbar} eBz_{mn} \int_0^t \cos\omega t \, e^{i\omega_{mn}t} dt$$

$$\omega_{mn} = \frac{E^{(m)} - E^{(n)}}{\hbar} > 0 \qquad \cos\omega t = \frac{e^{i\omega t} + e^{-i\omega t}}{2}$$

($\hbar\omega_{mn}$ between levels m and n)

this term only important when $\omega \approx \omega_{mn}$ then

$$a_m(t) \approx -\frac{ieB}{2\hbar} z_{mn} \int_0^t e^{i(\omega_{mn}-\omega)t} dt =$$

$$= +\frac{eB}{2\hbar} z_{mn} \frac{e^{-i(\omega-\omega_{mn})t} - 1}{\omega - \omega_{mn}}$$

(3) $$|a_m(t)|^2 = \frac{e^2B^2}{\hbar^2} |z_{mn}|^2 \frac{\sin^2 \frac{t}{2}(\omega-\omega_{mn})}{(\omega-\omega_{mn})^2}$$

Comments on resonance

Light intensity $= \frac{cB^2}{8\pi}$

Absorption from continuum overlapping ω_{mn}

(4) $$\frac{cB^2}{8\pi} = \frac{dI}{d\omega} d\omega$$ Substitute in (3), then $\int d\omega$ use $\int \frac{\sin^2 \alpha x}{x^2} dx = \pi\alpha$

$$|a_m|^2 = t \times \frac{4\pi^2 e^2}{c\hbar^2} |z_{mn}|^2 \frac{dI}{d\omega}$$

ω = ang. frequency not solid angle!

(5) $$\boxed{\text{Rate of absorption} = \frac{4\pi^2 e^2}{c\hbar^2} |z_{mn}|^2 \frac{dI}{d\omega}}$$

For isotropic radiation of volume energy density $u(\omega)\,d\omega$

factor 1/3 from averaging over directions of polarization

(6) $$\text{Rate of absorption} = \frac{4\pi^2 e^2}{3\hbar^2} |\vec{x}_{mn}|^2 u(\omega_{mn})$$

Relationship between emission & absorption could be derived from quantum electrodynamics — However simpler to use Einsteins A & B method

Rate of $n \to m$ $\quad B\,u(\omega)\,N(n) =$

[diagram: levels m (upper) and n (lower); upward arrow labelled BF; downward arrow labelled $A + CF$]

This B is a coefficient. Has nothing to do with B of page 1

From (6)

(7) $$B = \frac{4\pi^2 e^2}{3\hbar^2}\left|\vec{x}_{mn}\right|^2$$

Rate of $m \to n$) $\quad [A + C\,u(\omega)]\,N(m)$

this is number of atoms in state (n) or (m)

forced transitions

Spontaneous transitions

For thermal equilibrium

(8) $$\frac{N(m)}{N(n)} = e^{-\frac{E^{(m)} - E^{(n)}}{kT}} = e^{-\frac{\hbar\omega_{mn}}{kT}}$$ Boltzmann distribution

At equilibrium: Rate $n \to m$ = Rate $m \to n$

(9) $$\frac{A}{B\,u(\omega)} + \frac{C}{B} = \frac{N_n}{N_m} = e^{\frac{\hbar\omega}{kT}}$$

Planck's law

(10) $$u = \frac{\hbar\omega^3/\pi^2c^3}{e^{\frac{\hbar\omega}{kT}} - 1}$$

$$\frac{\pi^2c^3}{\hbar\omega^3}\frac{A}{B}\left(e^{\frac{\hbar\omega}{kT}} - 1\right) + \frac{C}{B} = e^{\frac{\hbar\omega}{kT}}$$

Must hold at all T's Therefore:

$$\frac{\pi^2c^3}{\hbar\omega^3}\frac{A}{B} = 1 \qquad \frac{C}{B} = 1$$

Einstein's relations

(11) $$\boxed{A = \frac{\hbar\omega^3}{\pi^2c^3}B\;;\quad C = B}$$ Then from (7)

(12) $$\boxed{\frac{1}{\tau} = A = \frac{4}{3}\frac{e^2\omega^3}{\hbar c^3}\left|\vec{x}_{mn}\right|^2}$$ for spontaneous transitions

(12) generalized to many particles by change

(13) $$e\vec{x} \to \sum e_i \vec{x}_i \quad \text{(sum to all particles)}$$

(14) $$\frac{1}{\tau} = \frac{4}{3}\frac{\omega^3}{\hbar c^3}\left|\sum e_i \langle m|\vec{x}_i|n\rangle\right|^2$$

Intensity of radiation proportional to square of matrix element of coordinates (for one electron) or of electric moment (13) for several charged particles.

<u>Discuss</u> – Limitations to validity of (12)

dimensions of atom $\ll \lambdabar$ of radiation

Quadrupole radiation

<u>Case of central forces – Selection rules</u> (See Sect 7)

Spherical harmonics identities

(15)
$$\sqrt{\frac{8\pi}{3}}\,Y_{11}\,Y_{l,m-1} = \sqrt{\frac{(l+m)(l+1+m)}{(2l+1)(2l+3)}}\,Y_{l+1,m} - \sqrt{\frac{(l-m)(l+1-m)}{(2l+1)(2l-1)}}\,Y_{l-1,m}$$

$$\sqrt{\frac{4\pi}{3}}\,Y_{10}\,Y_{l,m} = \sqrt{\frac{(l+1)^2-m^2}{(2l+1)(2l+3)}}\,Y_{l+1,m} + \sqrt{\frac{l^2-m^2}{(2l+1)(2l-1)}}\,Y_{l-1,m}$$

$$\sqrt{\frac{8\pi}{3}}\,Y_{1,-1}\,Y_{l,m+1} = \sqrt{\frac{(l-m)(l+1-m)}{(2l+1)(2l+3)}}\,Y_{l+1,m} - \sqrt{\frac{(l+m)(l+1+m)}{(2l+1)(2l-1)}}\,Y_{l-1,m}$$

(16) $$\sqrt{\frac{8\pi}{3}}\,Y_{11} = -\sin\vartheta\, e^{i\varphi}\,; \quad \sqrt{\frac{4\pi}{3}}\,Y_{10} = \cos\vartheta\,; \quad \sqrt{\frac{8\pi}{3}}\,Y_{1,-1} = \sin\vartheta\, e^{-i\varphi}$$

Follows: The matrix elements of the coordinates vanish unless

(17) $$l' = l \pm 1 \quad \text{and} \quad m' = m \pm 1 \text{ or } m$$ (Selection rules)

Matrix elements

$$(18)\begin{cases} \langle \overset{n'}{l+1}, m+1 | x+iy | \overset{n}{l}, m \rangle = -\mathcal{J}\sqrt{\dfrac{(l+m+2)(l+1+m)}{(2l+1)(2l+3)}} \\ \langle \overset{n'}{l+1}, m+1 | x-iy | \overset{n}{l}, m \rangle = 0 \\ \langle n', l+1, m | z | n, l, m \rangle = \mathcal{J}\sqrt{\dfrac{(l+1)^2-m^2}{(2l+1)(2l+3)}} \\ \langle n', l+1, m-1 | x+iy | n, l, m \rangle = 0 \\ \langle n', l+1, m-1 | x-iy | n, l, m \rangle = \mathcal{J}\sqrt{\dfrac{(l+1-m)(l+2-m)}{(2l+1)(2l+3)}} \end{cases}$$

$$(19)\qquad \mathcal{J} = \int_0^\infty R_{nl}(r)\, R_{n',l+1}(r)\, r^3\, dr$$

Derive

$$(20)\begin{cases} |\langle n', l+1, m+1 | \vec{x} | n, l, m \rangle|^2 + |\langle n', l+1, m | \vec{x} | n\, l\, m \rangle|^2 + \\ + |\langle n', l+1, m-1 | \vec{x} | n, l, m \rangle|^2 = \dfrac{l+1}{2l+1}\mathcal{J}^2 \quad (\text{indep. of } m) \end{cases}$$

(21) Therefore: rate of transition

$$(n, l, m) \rightarrow (n', l+1, \text{any } m')$$

$$= \frac{4}{3}\frac{e^2\omega^3}{\hbar c^3}\frac{l+1}{2l+1}\mathcal{J}^2$$

(Comments on independence of m)

Similarly

$$(22)\begin{cases} \text{Rate}(n, l, m \rightarrow n', l-1, \text{any } m) = \\ = \dfrac{4}{3}\dfrac{e^2\omega^3}{\hbar c^3}\dfrac{l}{2l-1}\left\{\displaystyle\int_0^\infty R_{nl}(r)\, R_{n',l-1}(r)\, r^3\, dr\right\}^2 \end{cases}$$

Example – Life time of 2p state of hydrogen

$$R_{1s}(r) = \frac{2}{a^{3/2}} e^{-r/a}; \quad R_{2p}(r) = \frac{1}{\sqrt{24a^3}} \frac{r}{a} e^{-r/2a}$$

$$J = \int R_{1s} R_{2p} r^3 dr = \frac{192\sqrt{2}}{243} a$$

$$\text{Rate}(2p \to 1s) = \frac{294912}{177147} \frac{e^2 \omega^3 a^2}{\hbar c^3} \qquad \omega = \frac{3}{4} \frac{me^4}{2\hbar^3}$$

$$= \frac{1152}{6561} \left(\frac{e^2}{\hbar c}\right)^3 \left(\frac{me^4}{2\hbar^3}\right) \qquad a = \frac{\hbar^2}{me^2}$$

$$= 1.41 \times 10^9 \text{ sec}^{-1}$$

$$\frac{e^2}{\hbar c} = \frac{1}{137} \qquad \frac{me^4}{2\hbar^3} = \frac{\text{Ryd}}{\hbar} = 2.067 \times 10^{16} \text{ sec}^{-1}$$

Topics for discussion

Permitted & forbidden lines

Metastable states

Generalization of selection rules

Irradiation by a linear oscillator

Sum rule & effective number of electrons

Polarization of emitted light

25 - Pauli theory of spin,

Int. degree of freedom - dicotomic variable - Operators on spin variable

(1) $$\begin{vmatrix} a_{11} & a_{12} \\ a_{21} & a_{22} \end{vmatrix}$$

Search for operators

(2) $$\sigma_x \,,\ \sigma_y \,,\ \sigma_z$$

Normalize them to e.v.'s ± 1. Then

(2) $$\sigma_x^2 = \sigma_y^2 = \sigma_z^2 = 1 = \begin{vmatrix} 1 & 0 \\ 0 & 1 \end{vmatrix}$$

Also

(3) $$(\alpha\sigma_x + \beta\sigma_y + \gamma\sigma_z)^2 = 1 \qquad \alpha, \beta, \gamma = \text{direction cosines}$$

(4) $$\rightarrow \sigma_x\sigma_y + \sigma_y\sigma_x = 0 \,, \ldots \quad \text{(anticommutation)}$$

Choose base for σ_z diagonal

(5) $$\sigma_z = \begin{vmatrix} 1 & 0 \\ 0 & -1 \end{vmatrix}$$

Ⓔ $\sigma_x = \begin{vmatrix} a & b \\ b^* & c \end{vmatrix}$ from $\sigma_x\sigma_z + \sigma_z\sigma_x = 1$ follows

$$\begin{vmatrix} a, & -b \\ b^*, & -c \end{vmatrix} + \begin{vmatrix} a & b \\ -b^* & -c \end{vmatrix} = 0 \longrightarrow \{ a = c = 0$$

$$\sigma_x = \begin{vmatrix} 0 & b \\ b^* & 0 \end{vmatrix} \qquad \sigma_x^2 = \begin{vmatrix} |b|^2 & 0 \\ 0 & |b|^2 \end{vmatrix} = \begin{vmatrix} 1 & 0 \\ 0 & 1 \end{vmatrix} \longrightarrow |b|^2 = 1$$

B $\sigma_x = \begin{vmatrix} 0 & e^{i\alpha} \\ e^{-i\alpha} & 0 \end{vmatrix}$ Dispose of phases of base vectors to make $\alpha = 1$. Then

(6) $$\sigma_x = \begin{vmatrix} 0 & 1 \\ 1 & 0 \end{vmatrix}$$

As above $\sigma_y = \begin{vmatrix} 0 & e^{i\beta} \\ e^{-i\beta} & 0 \end{vmatrix}$, From $\sigma_x\sigma_y + \sigma_y\sigma_x = 0$, find $e^{i\beta} + e^{-i\beta} = 0$ or $e^{i\beta} = \pm i$

σ_y = either $\begin{vmatrix} 0 & i \\ -i & 0 \end{vmatrix}$ or $\begin{vmatrix} 0 & -i \\ i & 0 \end{vmatrix}$

Eliminate first choice. Because

Ⓔ $\sigma_z = \begin{vmatrix} 1 & 0 \\ 0 & -1 \end{vmatrix}$ $\sigma_x = \begin{vmatrix} 0 & 1 \\ 1 & 0 \end{vmatrix}$ $\sigma_y = \begin{vmatrix} 0 & i \\ -i & 0 \end{vmatrix}$

First consider in place of $\vec{\sigma}$, $-\vec{\sigma}$ or $\vec{\sigma} \to -\vec{\sigma}$

$$\sigma_z = \begin{vmatrix} -1 & 0 \\ 0 & 1 \end{vmatrix} \quad \sigma_x = \begin{vmatrix} 0 & -1 \\ -1 & 0 \end{vmatrix} \quad \sigma_y = \begin{vmatrix} 0 & -i \\ i & 0 \end{vmatrix}$$

Then unitary transf. $T = \sigma_y$ transforms to standard form of Pauli spin operators

(7) $$\sigma_x = \begin{vmatrix} 0 & 1 \\ 1 & 0 \end{vmatrix}; \quad \sigma_y = \begin{vmatrix} 0 & -i \\ i & 0 \end{vmatrix}; \quad \sigma_z = \begin{vmatrix} 1 & 0 \\ 0 & -1 \end{vmatrix}$$

Check from (7)

(8) $$\sigma_x^2 = \sigma_y^2 = \sigma_z^2 = 1 \qquad \vec{\sigma}^2 = \sigma_x^2 + \sigma_y^2 + \sigma_z^2 = 3$$

(9) $$\sigma_x \sigma_y + \sigma_y \sigma_x = 0; \quad \sigma_y \sigma_z + \sigma_z \sigma_y = 0; \quad \sigma_z \sigma_x + \sigma_x \sigma_z = 0$$

(10) $$\sigma_x \sigma_y = i\sigma_z; \quad \sigma_y \sigma_z = i\sigma_x; \quad \sigma_z \sigma_x = i\sigma_y$$

(11) $$[\sigma_x, \sigma_y] = 2i\sigma_z; \quad [\sigma_y, \sigma_z] = 2i\sigma_x; \quad [\sigma_z, \sigma_x] = 2i\sigma_y$$

or

(12) $$\vec{\sigma} \times \vec{\sigma} = 2i\vec{\sigma}$$

Consider vector

(13) $$\vec{s} = \frac{\hbar}{2}\vec{\sigma}$$ Then,

(14) $$\vec{s} \times \vec{s} = i\hbar\vec{s}$$

Identical to exch. rules (18-(5)) or (20-(26)) of ang. mom. vectors. Therefore $\vec{s} = \frac{\hbar}{2}\vec{\sigma}$ = intrinsic ang. mom of electron.

(15) E.v. of s_x, s_y, s_z are $\pm \frac{\hbar}{2}$

Also $\vec{s}^2 = s_x^2 + s_y^2 + s_z^2 = \frac{\hbar^2}{4}\vec{\sigma}^2 = \frac{3}{4}\hbar^2 = \hbar^2 \frac{1}{2}\times(\frac{1}{2}+1)$

Both mean: Spin angular momentum $= \hbar/2$

Magnetic moment. Zeeman effect requires that spin carries a magn. moment

(16) $$\vec{\mu} = \mu_0 \vec{\sigma} \qquad \mu_0 = \frac{e\hbar}{2mc} = \text{Bohr magneton}$$

Same conclusion from Dirac relativistic theory of electron. Schwinger (1948) computed radiative correction

(17) $$\mu_0 = \frac{e\hbar}{2mc}\left(1 + \frac{1}{2\pi}\frac{e^2}{\hbar c}\right) = \frac{e\hbar}{2mc} \times 1.00116$$

in better agreement with expt.

When electron moves in ext magn. field B (∥ to z axis) add to Hamiltonian (21-(27)) the term

(18) $$-B\mu_0\sigma_z = -B\frac{e\hbar}{2mc}\sigma_z$$

Observe

$$\frac{\text{mag. moment}}{\text{ang. momentum}/\hbar} = \begin{cases} \mu_0 & \text{for orbital motion} \\ 2\mu_0 & \text{for spin} \end{cases}$$

Topics for discussion — Motion of an isolated spin vector in a constant or variable magnetic field. Meaning of direction of spin vector

26 – Electron in central field.

(1) Potential $= -e\,V(r)$

Spin orbit interaction (Classical)

$$E = -\frac{dV}{dr}$$

Apparent mag. field for electron

$$(2)\quad \begin{cases} \approx -\frac{1}{c}\vec{v}\times\vec{E} & \vec{E} = -\frac{dV}{dr}\frac{\vec{x}}{r} \\ = -\frac{1}{c}\frac{1}{r}\frac{dV}{dr}\vec{x}\times\vec{v} = -\frac{1}{mc}\frac{1}{r}V'(r)\,\vec{M} = -\frac{\hbar}{mc}\frac{V'(r)}{r}\vec{L} \end{cases}$$

(3) $\begin{cases} \vec{M} = \text{orb. ang. momentum} = \hbar\vec{L} \\ \text{Mag. moment of electron} = \mu_0\vec{\sigma} = \frac{e\hbar}{2mc}\vec{\sigma} \end{cases}$

Mutual energy of intrinsic mag. mom and apparent field

$$(4)\quad -\frac{V'(r)}{r}\frac{\hbar\mu_0}{mc}\left(\vec{L}\cdot\vec{\sigma}\right) = \frac{-e\hbar^2\,V'(r)}{2m^2c^2r}\left(\vec{L}\cdot\vec{\sigma}\right)$$

(minus sign because electron negative)

Thomas correction. Is a relativistic term that cancels half of (4) – Also from completely relativistic Dirac theory. In conclusion:

Spin orbit interaction adopted

$$(5)\quad -\frac{\hbar\mu_0}{2mc}\frac{V'(r)}{r}\left(\vec{L}\cdot\vec{\sigma}\right) = -\frac{e\hbar^2}{4m^2c^2}\frac{V'(r)}{r}\left(\vec{L}\cdot\vec{\sigma}\right)$$

Hamiltonian of electron

$$(6)\quad H = \frac{1}{2m}p^2 - eV(r) - \frac{e\hbar^2}{4m^2c^2}\frac{V'(r)}{r}\left(\vec{L}\cdot\vec{\sigma}\right)$$

Put

(7) $\vec{S} = \frac{\vec{\sigma}}{2}$ (this = intrinsic spin ang. mom. in unit $\hbar$)

(8) $$\begin{cases} H = \frac{1}{2m} p^2 - e\,V(r) - \frac{e\hbar^2 V'(r)}{2m^2 c^2 r}(\vec{L}\cdot\vec{S}) \\ \quad = H_1 + H_2\,(\vec{L}\cdot\vec{S}) \end{cases} \qquad H_1 = \frac{1}{2m}p^2 - e\,V(r)$$

Introduce also: $\qquad H_2 = -\frac{e\hbar^2}{2m^2c^2}\frac{V'(r)}{r}$

(9) $\vec{J} = \vec{L} + \vec{S}$ = tot. ang. mom. in $\hbar$ units.

List of commutation properties:

(10) $$\begin{cases} \vec{L}\times\vec{L} = i\vec{L}\ ; \quad \vec{S}\times\vec{S} = i\vec{S} \\ [L_x, L_y] = i\,L_z \text{ \& similar } [L_x, L^2] = 0, \ldots \\ [S_x, S_y] = i\,S_z \text{ \& " } \quad [S_x, S^2] = 0, \ldots \end{cases}$$

(11) $\{[L_x, S_x] = 0 \quad [L_x, S_y] = 0$ and similar

(12) $\{ \quad S^2 = \frac{3}{4}$

Follows from (10) (11) (9)

(13) $\vec{J}\times\vec{J} = i\vec{J}$ or $[J_x, J_y] = i\,J_z$ & similar

$\vec{J}$ behaves like an ang. mom. vector. From (13)

(14) $[J_x, J^2] = 0$, and similar

(15) $\begin{cases}\text{All components of } \vec{L}, \vec{S}, \vec{J} \text{ and also } L^2, S^2 = \frac{3}{4}, J^2 \\ \text{commute with } H_1, H_2.\end{cases}$

(16) $[(\vec{L}\cdot\vec{S}), J_x] = 0$

Proof: $[(L_xS_x + L_yS_y + L_zS_z), (L_x + S_x)] = [L_yL_x]S_y + [L_zL_x]S_z +$
$+ L_y[S_yS_x] + L_z[S_zS_x] = -i\,L_zS_y + i\,L_yS_z - i\,L_yS_z + i\,L_zS_y = 0$

$$(16)\begin{cases} [(\vec{L}\cdot\vec{S}), J^2] = 0 \\ [(\vec{L}\cdot\vec{S}), L^2] = 0 \\ [(\vec{L}\cdot\vec{S}), S^2] = 0 \end{cases}$$

Therefore

$$(17)\qquad [H, J^2] = [H, L^2] = [H, S^2] = 0$$

Also

$$(18)\qquad [H, (L\cdot S)] = 0$$

$$(19)\qquad [H, J_x] = [H, J_y] = [H, J_z] = 0$$

$$(20)\qquad J^2 = L^2 + S^2 + 2(L\cdot S)$$

Hence

$$(21)\qquad [J^2, L^2] = [J^2, S^2] = 0$$

$$(22)\qquad [J_z, L^2] = [J_z, S^2] = [J_z, J^2] = 0$$

First characterize state by making diagonal following intercommuting quantities

$$23\ \begin{cases} H_1,\ H_2,\ L^2 = l(l+1),\ S^2 = \frac{3}{4},\ L_z = m_l,\ S_z = m_s \\ m_l = l, l-1, \ldots, -l+1, -l \qquad J_z = m_l + m_s = m \\ m_s = \pm 1/2 \qquad l - \frac{1}{2} \leq J_z \leq l + \frac{1}{2} \end{cases}$$

H in general <u>not</u> diagonal because $(L\cdot S)$ does not commute with L_z or S_z. But $[(L\cdot S), J_z] = 0$

Therefore $(L\cdot S)$ mixes states of same $J_z = m$ and different L_z, S_z. Two such states:

(24)

$L_z = m - \frac{1}{2}$, $S_z = \frac{1}{2}$ state $|m-\frac{1}{2}, \frac{1}{2}\rangle$

and $L_z = m + \frac{1}{2}$ $S_z = -\frac{1}{2}$ state $|m+\frac{1}{2}, \frac{1}{2}\rangle$

$$|m-\tfrac{1}{2}, \tfrac{1}{2}\rangle = \psi_{m-\frac{1}{2}, \frac{1}{2}} = f(r)\, Y_{l, m-\frac{1}{2}} \begin{vmatrix} 1 \\ 0 \end{vmatrix}$$

$$|m+\tfrac{1}{2}, \tfrac{-1}{2}\rangle = \psi_{m+\frac{1}{2}, \frac{1}{2}} = f(r)\, Y_{l, m+\frac{1}{2}} \begin{vmatrix} 0 \\ 1 \end{vmatrix}$$

Find from Lects (18 especially (17) (18)) and

(25) Lect (25) use $S_x + iS_y = \begin{vmatrix} 0 & 1 \\ 0 & 0 \end{vmatrix}$ $S_x - iS_y = \begin{vmatrix} 0 & 0 \\ 1 & 0 \end{vmatrix}$ $S_z = \begin{vmatrix} 1 & 0 \\ 0 & -1 \end{vmatrix}$

(26) $$(L\cdot S) = \tfrac{1}{2}(L_x + iL_y)(S_x - iS_y) + \tfrac{1}{2}(L_x - iL_y)(S_x + iS_y) + L_z S_z$$

(27)

$$(L_x + iL_y)\, Y_{l, m-\frac{1}{2}} = \sqrt{(l+\tfrac{1}{2})^2 - m^2}\; Y_{l, m+\frac{1}{2}}$$

$$(L_x - iL_y)\, Y_{l, m+\frac{1}{2}} = \sqrt{(l+\tfrac{1}{2})^2 - m^2}\; Y_{l, m-\frac{1}{2}}$$

($m \pm \frac{1}{2}$ = integral number)

(28)

$$(S_x + iS_y)\begin{vmatrix} 1 \\ 0 \end{vmatrix} = 0 \qquad (S_x + iS_y)\begin{vmatrix} 0 \\ 1 \end{vmatrix} = \begin{vmatrix} 1 \\ 0 \end{vmatrix}$$

$$(S_x - iS_y)\begin{vmatrix} 0 \\ 1 \end{vmatrix} = 0 \qquad (S_x - iS_y)\begin{vmatrix} 1 \\ 0 \end{vmatrix} = \begin{vmatrix} 0 \\ 1 \end{vmatrix}$$

Find

(29)

$$(L\cdot S)\,|m-\tfrac{1}{2}, \tfrac{1}{2}\rangle = \tfrac{1}{2}(m-\tfrac{1}{2})\,|m-\tfrac{1}{2}, \tfrac{1}{2}\rangle + \tfrac{1}{2}\sqrt{(l+\tfrac{1}{2})^2 - m^2}\;|m+\tfrac{1}{2}, \tfrac{-1}{2}\rangle$$

$$(L\cdot S)\,|m+\tfrac{1}{2}, -\tfrac{1}{2}\rangle = \tfrac{1}{2}\sqrt{(l+\tfrac{1}{2})^2 - m^2}\;|m-\tfrac{1}{2}, \tfrac{1}{2}\rangle - \tfrac{1}{2}(m+\tfrac{1}{2})\,|m+\tfrac{1}{2}, \tfrac{-1}{2}\rangle$$

(30) $$(L\cdot S) = \begin{Vmatrix} \frac{1}{2}(m-\frac{1}{2}), & \frac{1}{2}\sqrt{(l+1/2)^2 - m^2} \\ \frac{1}{2}\sqrt{(l+1/2)^2 - m^2}, & -\frac{1}{2}(m+\frac{1}{2}) \end{Vmatrix}$$

e.v's of $(\vec{L}\cdot\vec{S})$ & corresp e.f's are

$$(31)\quad \begin{cases} \vec{L}\cdot\vec{S} = \frac{1}{2}l \text{ with e.f (normalized)} \\ \sqrt{\frac{1}{2}+\frac{m}{2l+1}}\left|m-\frac{1}{2},\frac{1}{2}\right\rangle + \sqrt{\frac{1}{2}-\frac{m}{2l+1}}\left|m+\frac{1}{2},\frac{-1}{2}\right\rangle \end{cases}$$

and

$$(32)\quad \begin{cases} \vec{L}\cdot\vec{S} = -\frac{1}{2}(l+1) \text{ with normalized e.f.} \\ -\sqrt{\frac{1}{2}-\frac{m}{2l+1}}\left|m-\frac{1}{2},\frac{1}{2}\right\rangle + \sqrt{\frac{1}{2}+\frac{m}{2l+1}}\left|m+\frac{1}{2},\frac{-1}{2}\right\rangle \end{cases}$$

E.v's of J^2 from (20) (31) (32)

$$(33)\quad \begin{cases} \text{for } L\cdot S = \frac{l}{2},\ J^2 = l(l+1)+\frac{3}{4}+l = (l+\frac{1}{2})(l+\frac{1}{2}+1) \\ s \parallel \text{ to } l \text{ or vector model},\ J = l+\frac{1}{2} \\ \vec{J}^2 = J(J+1).\ \text{E.f is (31).} \end{cases}$$

$$(34)\quad \begin{cases} \text{for } L\cdot S = -\frac{1}{2}(l+1),\ J^2 = l(l+1)+\frac{3}{4}-l-1 = (l-\frac{1}{2})(l+\frac{1}{2}) \\ \text{Spin antiparallel to } l,\ J = l-\frac{1}{2} \\ \text{E.f is (32)}, \quad \vec{J}^2 = J(J+1) = l^2-\frac{1}{4} \end{cases}$$

<u>Doublet splitting of energy levels.</u> From (8)

$$(35)\quad -\frac{e\hbar^2}{2m^2c^2}\frac{V'(r)}{r}(L\cdot S)$$

treated as perturbation, yields energy perturbation

this, usually, positive; R_l = radial wave function

$$(36)\quad \delta E = \left\{\frac{e\hbar^2}{2m^2c^2}\right\}\left(\int\left\{V'(r)\right\}R_l^2(r)\,r\,dr\right)\times\begin{cases} l/2 & \text{for } J=l+\frac{1}{2} \\ \text{or} & \\ -(l+1)/2 & \text{for } J=l-\frac{1}{2} \end{cases}$$

Doublet spectrum (Typical case Alkali atoms)

(s) $l=0$ J: — 1/2; — 1/2

(p) $l=1$ J: = 3/2, 1/2

(d) $l=2$ J: = 5/2, 3/2

Notation: $s_{1/2}$, $p_{1/2}$, $p_{3/2}$, $d_{3/2}$, $d_{5/2}$

3/2, 1/2 → 1/2: D lines of sodium $\lambda = 5890$ Å and $\lambda = 5896$ Å

Case of $n=2$ levels of hydrogen. From Lect. 8

$$E = -\frac{me^4}{2\hbar^2 \times 2^2} \text{ for } 2s + 2p \text{ levels.}$$

Spin perturbation (36) ($\delta_1 E$)

(37) $\delta_1 E(2s) = 0 \qquad \delta_1 E(2p) = \frac{e^2\hbar^2}{48 m^2 c^2}\frac{1}{a^3}\begin{cases} 1/2 \\ -1 \end{cases}$

$2s_{1/2}$ —— $2p_{3/2}$ —— $\propto 1/2$; $2p_{1/2}$ —— $\propto -1$

Use $R_{2p} = \frac{r e^{-r/2a}}{\sqrt{24 a^5}}$ and $V = \frac{e}{r}$ in (36)

Relativity perturbation ($\delta_2 E$)

(38) $\text{kin. energy} = \sqrt{m^2c^4 + c^2p^2} - mc^2 = \frac{p^2}{2m} - \frac{p^4}{8m^3c^2} + \ldots$

(39) $\text{Perturbation} = -\frac{1}{8m^3c^2}\, p^4 = -\frac{\hbar^4}{8m^3c^2}(\nabla^2)^2$

(40) One finds using first approx. perturbation theory

$\delta_2 E(2s) = -\frac{5}{128}\frac{e^8 m}{\hbar^4 c^2} \qquad \delta_2 E(2p) = -\frac{7}{384}\frac{e^8 m}{\hbar^4 c^2}$

(See for general formulas: Schiff p. 325, 326)

$$\delta_1(E_{2s}) + \delta_2(E_{2s}) = -\frac{5}{128}\frac{e^8 m}{\hbar^4 c^2}$$ ← !!

$$\delta_1(E_{2p_{1/2}}) + \delta_2(E_{2p_{1/2}}) = \left(-\frac{1}{48} - \frac{7}{384}\right)\frac{e^8 m}{\hbar^4 c^2} = -\frac{5}{128}\frac{e^8 m}{\hbar^4 c^2}$$

$$\delta_1(E_{2p_{3/2}}) + \delta_2(E_{2p_{3/2}}) = \left(\frac{1}{96} - \frac{7}{384}\right)\frac{e^8 m}{\hbar^4 c^2} = -\frac{1}{128}\frac{e^8 m}{\hbar^4 c^2}$$

unperturbed: $2s, 2p_{1/2}, 2p_{3/2}$

δ_1: $2p_{3/2}$, $2s$, $2p_{1/2}$

$\delta_1 + \delta_2$: $2p_{3/2}$; .365 cm^{-1}; $2s, 2p_{1/2}$

$\delta_1 + \delta_2 +$ Lamb: $2p_{3/2}$; $2s$; .035 cm^{-1}; $2p_{1/2}$

Qualitative comments on Lamb shift:

Bethe formula for Lamb shift of ns-levels

$$\frac{8}{3\pi n^3}\frac{me^4}{2\hbar^2}\left(\frac{e^2}{\hbar c}\right)^3 \overline{\ln\frac{mc^2}{|E_n - E_s|}} + \text{higher order corrections}$$

27 - Anomalous Zeeman effect.

To prec. case add mag. field B ∥ to z

Magn. energy

(1) $$B\mu_0 (L_z + 2S_z)$$

Unpert. hamiltonian

(2) $$H_1 = \frac{p^2}{2m} - eV(r)$$

Perturbation

(3) $$\mathcal{H} = \frac{e\hbar^2}{2m^2c^2} \frac{-V'(r)}{r} (\vec{L} \cdot \vec{S}) + B\mu_0 (L_z + 2S_z)$$

(4) Observe L^2, $S^2 = \frac{3}{4}$, $m = L_z + S_z$ commute with $\mathcal{H}$, ~~all these~~

(5) Unperturbed problem has 2l-fold deg.
Unpert. e.f's

$$R_l(r)\, Y_{lm}(\theta, \varphi) \times \text{spin} \begin{pmatrix}\text{up} \\ \text{or} \\ \text{down}\end{pmatrix}$$

Clebsch-Gordan Coeff of expression (26-(36))

(6) $$k = \frac{e\hbar^2}{2m^2c^2} \int (-V'(r))\, R_l^2(r)\, r\, dr$$

Pert. matrix mixes states (26-(24)) see also (26-(39))

(7) $$\frac{k}{2} \begin{vmatrix} m - \frac{1}{2} & \sqrt{(l+\frac{1}{2})^2 - m^2} \\ \sqrt{(l+\frac{1}{2})^2 - m^2}, & -m - \frac{1}{2} \end{vmatrix} + B\mu_0 \begin{vmatrix} m + \frac{1}{2} & 0 \\ 0 & m - \frac{1}{2} \end{vmatrix}$$

Find eigenvalues as roots of

(8) $x^2+\left(\frac{k}{2}-2B\mu_0 m\right)x+\left(m^2-\frac{1}{4}\right)B^2\mu_0^2-B\mu_0 km-\frac{k^2}{4}l(l+1)=0$

(9) $\delta E=-\frac{k}{4}+B\mu_0 m\pm\frac{1}{2}\sqrt{k^2\left(l+\frac{1}{2}\right)^2+2B\mu_0 km+B^2\mu^2}$

(9) valid for $|m|\leq l-\frac{1}{2}$,
for $m=\pm(l+\frac{1}{2})$, $\delta E=\frac{k}{2}l\pm B\mu_0(l+1)$

For $B\mu_0\ll k$

(10) $$\delta E=\begin{cases}\frac{k}{2}l+B\mu_0 m\,\frac{2l+2}{2l+1} & -l-\frac{1}{2}\leq m\leq l+\frac{1}{2}\\ -\frac{k}{2}(l+1)+B\mu_0 m\,\frac{2l}{2l+1} & -l+\frac{1}{2}\leq m\leq l-\frac{1}{2}\end{cases}$$

For $B\mu_0\gg k$

(11) $$\delta E=\begin{cases}B\mu_0\left(m+\frac{1}{2}\right)\\ B\mu_0\left(m-\frac{1}{2}\right)\end{cases}$$

For $l=1$

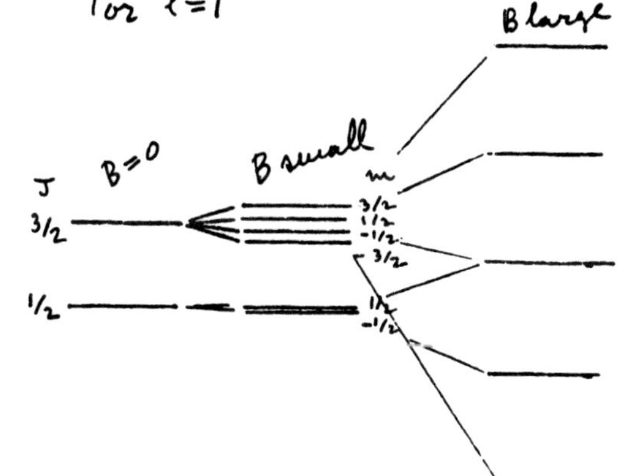

28 – Addition of ang. momentum vectors.

(1) $\vec{L}$, $\vec{S}$, $\vec{L}+\vec{S}=\vec{J}$

Assume

(2) $[\vec{L},\vec{S}]=0$

L orbital
S spin
J total

(3) $\vec{L}\times\vec{L}=i\vec{L}$, $\vec{S}\times\vec{S}=i\vec{S}$ ($\hbar=1$)

Follows

(4) $\vec{J}\times\vec{J}=i\vec{J}$

Two intercommuting sets of operators:

(5) Set a) L^2, S^2, L_z, S_z

(6) Set b) L^2, S^2, J^2, J_z

First: operators (a) diagonal

(7)
$$\begin{cases} L^2=l(l+1) & S^2=s(s+1) \\ L_z=\lambda & S_z=\mu. \\ \lambda=-l,-l+1,\dots,l-2,\ l-1,\ l & \\ \mu=-s,-s+1,\dots,s-1,s & \end{cases}$$

l, s are integrals or half odd numbers. When l is the result of orbital ang. mom. l is integral. When s is the resultant spin s is integral for even number of electrons, half odd for odd number of electrons.

An eigenvector for (7)

(8) $|L_z=\lambda, S_z=\mu\rangle$
or briefly $|\lambda,\mu\rangle$. $(2l+1)\times(2s+1)$ such vectors

Representation with vectors $|\lambda,\mu\rangle$ transformed now to a new one with set (b)

Operators for (6) diagonal

(9)
$$\begin{cases} L^2 = l(l+1) \qquad S^2 = s(s+1) \\ J^2 = j(j+1) \qquad J_z = L_z + S_z = m \\ j = \text{integer or half odd} \\ m = -j, -j+1, \ldots, j-2, j-1, j \end{cases}$$

Eigenvectors for (9)

(10)
$$\begin{cases} |J^2 = j(j+1); J_z = m\rangle \quad \text{or briefly} \\ |j, m\rangle \end{cases}$$

Question: Given l, s what are the possible values of j?

(11) Vector model rule $\quad j = l+s, l+s-1, \ldots, |l-s|$

Hint of proof:

(12) $m = \lambda + \mu \qquad \lambda \leq l, \ \mu \leq s$

$m \leq l+s$ Therefore $j_{max} = l+s$

Observe

(13) $\quad |\lambda = l, \mu = s\rangle = |j = l+s, m = l+s\rangle$

(14) Apply to (13) $J_- = J_x - iJ_y = L_x - iL_y + S_x - iS_y$ to obtain successively $|j=l+s, m=l+s\rangle$, $|j=l+s, m=l+s-1\rangle, \ldots, |j=l+s, m=-j\rangle$

These are $2(l+s)+1$ eigenvectors of type (10)

$m = \ell + s - 1$ possible in two ways

(15) $|\lambda = \ell - 1, \mu = s\rangle$ or $|\lambda = \ell, \mu = s - 1\rangle$

One linear comb. already under (14), other lin. comb. has

(16) $\left\{ \begin{array}{l} |j = \ell + s - 1, m = j\rangle \\ | \quad '' \quad j-1\rangle \\ | \quad '' \quad j-2\rangle \\ \vdots \quad\quad \vdots \\ \quad\quad -j\rangle \end{array} \right\}$ ← apply J_- to form $2(\ell+s)-1$ eigen vector of type (10)

and so forth.

Clebsch - Gordan coefficients

(17) $\left\{ \begin{array}{l} \langle \lambda, \mu | j, m\rangle = 0 \quad \text{for } \lambda + \mu \neq m \\ \langle \lambda, m - \lambda | j, m\rangle \text{ obtained by following} \end{array} \right.$

above procedure — General formulas are extremely complicated. Important special cases: $s = 1/2$ (See (26- (31)(32))

(18)

$s = 1/2$

	$\ell_z = m - \frac{1}{2}$, $s_z = 1/2$	$\ell_z = m + 1/2$, $s_z = -1/2$
$j = \ell + \frac{1}{2}$	$\sqrt{\frac{1}{2} + \frac{m}{2\ell+1}}$	$\sqrt{\frac{1}{2} - \frac{m}{2\ell+1}}$
$j = \ell - \frac{1}{2}$	$-\sqrt{\frac{1}{2} - \frac{m}{2\ell+1}}$	$\sqrt{\frac{1}{2} + \frac{m}{2\ell+1}}$

S = 1

(19)

	$\ell_z = m-1$, $s_z = 1$	$\ell_z = m$, $s_z = 0$	$\ell_z = m+1$, $s_z = -1$
$j=\ell+1$	$\sqrt{\frac{(\ell+m)(\ell+m+1)}{(2\ell+1)(2\ell+2)}}$	$\sqrt{\frac{(\ell-m+1)(\ell+m+1)}{(2\ell+1)(\ell+1)}}$	$\sqrt{\frac{(\ell-m)(\ell-m+1)}{(2\ell+1)(2\ell+2)}}$
$j=\ell$	$-\sqrt{\frac{(\ell+m)(\ell-m+1)}{2\ell(\ell+1)}}$	$\frac{m}{\sqrt{\ell(\ell+1)}}$	$\sqrt{\frac{(\ell-m)(\ell+m+1)}{2\ell(\ell+1)}}$
$j=\ell-1$	$\sqrt{\frac{(\ell-m)(\ell-m+1)}{2\ell(2\ell+1)}}$	$-\sqrt{\frac{(\ell-m)(\ell+m)}{\ell(2\ell+1)}}$	$\sqrt{\frac{(\ell+m+1)(\ell+m)}{2\ell(2\ell+1)}}$

More similar formulas in Condon & Shortley

Value of $\vec{L}\cdot\vec{S}$

(20) $$\vec{L}\cdot\vec{S} = \frac{1}{2}\{j(j+1) - \ell(\ell+1) - s(s+1)\}$$

Because $\vec{L} + \vec{S} = \vec{J}$

$$\vec{J}^2 = \vec{L}^2 + \vec{S}^2 + 2\,\vec{L}\cdot\vec{S}$$

Observe: (20) independent of <u>m</u>! more general

· <u>Theorem</u>: Classify e.f's by

(21) $$|n, j, m\rangle$$

Let A a rotation invariant operator.
(Means $[A, \vec{J}] = 0$). Then:

85 copies
28-5

(22) $\langle n', j', m' | A | n, j, m \rangle = \delta_{j,j'} \delta_{mm'} f(n, n', j)$

Comments & connection with Wigner theorem p. 20-4

Theorems on matrix elements of a vector operator $\vec{A}$

(23) $\langle n' j', m' | \vec{A} | n, j, m \rangle = 0$ except when
$j' = j+1, j, j-1$
$m' = m+1, m, m-1$

also
$\langle n', 0, 0 | \vec{A} | n, 0, 0 \rangle = 0$

Comments on selection rules for optical transitions

(24) Permitted transitions: $j \to j+1, j, j-1$ $\quad m \to m+1, m, m-1$

$j=0 \to j=0$ forbidden

(25) Selection rule for parity: for permitted transitions, change of parity.

(This is because electric moment is a polar vector)

Discuss: selection rules for electric quadrupole, magnetic dipole, etc..

(26) The matrix elements of the components of a vector are expressed as the product of a function
$f(n, n', j, j')$
times certain expressions that depend on j, j', m, m', and the component chosen.

Only different from zero

$$\langle m+1 | X+iY | m\rangle, \langle m | Z | m\rangle, \langle m-1 | X-iY | m\rangle$$

(explain)

X, Y, Z = components of $\vec{A}$

(27)
Transitions $j \to j+1$

$$\langle m+1 | X+iY | m\rangle \propto -\sqrt{(j+m+1)(j+m+2)}$$

$$\langle m | Z | m\rangle \propto \sqrt{(j-m+1)(j+m+1)}$$

$$\langle m-1 | X-iY | m\rangle \propto \sqrt{(j-m+1)(j-m+2)}$$

(28)
Transitions $j \to j$

$$\langle m+1 | X+iY | m\rangle \propto \sqrt{(j+m+1)(j-m)}$$

$$\langle m | Z | m\rangle \propto m$$

$$\langle m-1 | X-iY | m\rangle \propto \sqrt{(j-m+1)(j+m)}$$

(29)
Transitions $j \to j-1$

$$\langle m+1 | X+iY | m\rangle \propto -\sqrt{(j-m-1)(j-m)}$$

$$\langle m | Z | m\rangle \propto -\sqrt{j^2-m^2}$$

$$\langle m-1 | X-iY | m\rangle \propto \sqrt{(j+m)(j+m-1)}$$

Warning. Proportionality coefficients are different for (27) (28) (29).

Observe: in all 3 cases above

30) $\sum_{m'} |\langle m' | X | m\rangle|^2 + |\langle m' | Y | m\rangle|^2 + |\langle m' | Z | m\rangle|^2$ is independent of m. Comments on equal life time of states with different $\underline{m}$.

29 -- Atomic multiplets

Qualitative discussion

(1) $H = H_1 + H_2 (\vec{L}\cdot\vec{S})$

H_1, H_2 commute with $\vec{L}$ and $\vec{S}$. Then

H commutes with $\vec{L}^2$, $\vec{S}^2$, $\vec{J}^2$, J_z

Use (28-(2.2))

(2) $$\vec{L}\cdot\vec{S} = \frac{1}{2}\{J(J+1) - L(L+1) - S(S+1)\}$$

(3) note change of notation to usual spectroscopic notation $\vec{L}$, $\vec{S}$, $\vec{J}$ are vector operators

L, S, J are numbers (integers or half odd)

(4) Then for fixed values of L, S

$|L-S| \leq J \leq L+S$ J by integral steps

For a set of levels with n, L, S fixed

(5) $$H = H_1 + \frac{1}{2} H_2 \{J(J+1) - L(L+1) - S(S+1)\}$$

Assume H_2 small, then perturbation theory with H_1 diagonal (together with $\vec{L}^2, \vec{S}^2, \vec{J}$).

For an isolated group of levels H_1 & H_2 behave like numbers $H_2 \to$ its mean value $H_1 \to$ its diagonal value.

There is in multiplet one distinct energy level of each J value. From (4) J takes $2S+1$ values for $S \leq L$ or $2L+1$ values for $S > L$. However always called $(2S+1)$-plet. $S=0$, singlet; $S=\frac{1}{2}$, doublet

$S=1$, triplet ; ...

(6) $\begin{cases} H_2 > 0 & \text{normal multiplet} \\ H_2 < 0 & \text{inverted multiplet} \end{cases}$

value of L, by letter $S, P, D \ldots$

Notation $^3D_1$ and similar $^{2S+1}L_J$

Normal D-triplet

$^3D_3$ ——

$3H_2$

$^3D_2$ ——

$2H_2$

$^3D_1$ ——

Note: <u>Interval rule</u> – The spacing between two levels of multiplet ~~with~~ number J and $J+1$ is $\propto J+1$

Each of the multiplet levels is $2J+1$ fold

Degeneracy removed by magn. field $B \parallel z$,

This adds to energy perturbation term

(7) $$H_3 = B\mu_0(L_z + 2S_z) = B\mu_0(J_z + S_z) = \\ = B\mu_0(m + S_z)$$

assume

(8) $$H_3 \ll H_2$$

Then first approx pert. theory. Observe

$$[H_3, J_z] = 0$$

therefore no mixing of $2J+1$ degenerate terms. Then

(9) $$\delta E_3 = \langle J, m | H_3 | J, m \rangle = \\ = B\mu_0(m + \langle J, m | S_z | J, m \rangle)$$

From (28-(28))

(10) $$\langle J,m|S_z|J,m\rangle = \frac{\langle J,J|S_z|J,J\rangle}{J} m$$

Also

(11) $$\langle J,J|S_z|J,J\rangle = \frac{S(S+1)+J(J+1)-L(L+1)}{2(J+1)}$$

Outline of proof: From $\vec{L} = \vec{J} - \vec{S}$

$$2\,\vec{J}\cdot\vec{S} = J(J+1) + S(J+1) - L(L+1)$$

$$2\,\vec{J}\cdot\vec{S} = 2J_zS_z + S_-J_+ + S_+J_- \qquad J_\pm = J_x \pm iJ_y$$

$$= 2(J_z+1)S_z + S_-J_+ + J_-S_+ \longleftarrow \text{use} \qquad S_\pm = S_x \pm iS_y$$

$$S_xS_y - S_yS_x = iS_z$$

Use $J_+|J,J\rangle = 0 \qquad \langle J,J|J_- = 0$

Find

$$\langle J,J|2\,\vec{J}\cdot\vec{S}|J,J\rangle = 2(J+1)\langle J,J|S_z|J,J\rangle$$, hence proof

Then

(12) $$\delta E_3 = B\mu_0 g\, m$$

(13) $$\begin{cases} g = 1 + \dfrac{J(J+1)+S(S+1)-L(L+1)}{2J(J+1)} & \text{(Landé } g\text{-factor)} \\ g = \dfrac{3}{2} + \dfrac{S(S+1)-L(L+1)}{2J(J+1)} \end{cases}$$

Compare with (27-(10)) for case $S = 1/2$

For discussion

(14) Limiting case
$$B\mu_0 \gg H_2$$
(Paschen Back effect)

Selection & polarization rules from (28-(27)(28)(29))
For permitted transitions

(15) $J \nearrow J+1, \; J \rightarrow J, \; J \searrow J-1$ ($J = 0 \rightarrow 0 = J$ forbidden)

(16)
- $m \rightarrow m$ polarized $\parallel$
- $m \rightarrow m+1$ polarized ↻ } both $\perp$
- $m \rightarrow m-1$ " ↺ }

also parity rule

(17)
even → odd
odd → even

weaker selection rules

(18)
$S \rightarrow S$ } especially for
$L \nearrow L+1, \; L \rightarrow L, \; L \searrow L-1$ } light elements

---

Topics for discussion.
General data on atomic structure, Screening
Pauli principle (as empirical rule)
Atomic shells (table on next page)
Spectra of alkali's, alkaline earths and earths. Spectral series. Spectra of ions.
Electrons & holes in a shell.
Hyperfine structure
multiplets

Electron Orbits of Atoms

	n=1 K	n=2 L		n=3 M			n=4 N				n=5 O					n=6 P						n=7 Q						
L =	0	0	1	0	1	2	0	1	2	3	0	1	2	3	4	0	1	2	3	4	5	0	1	2	3	4	5	6
1 H	1																											
2 He	2																											
3 Li	2	1																										
4 Be	2	2																										
5 B	2	2	1																									
10 Ne	2	2	6																									
11 Na	2	2	6	1																								
12 Mg	2	2	6	2																								
13 Al	2	2	6	2	1																							
18 A	2	2	6	2	6																							
19 K	2	2	6	2	6		1																					
20 Ca	2	2	6	2	6		2																					
29 Cu	2	2	6	2	6	10	1																					
30 Zn	2	2	6	2	6	10	2																					
31 Ga	2	2	6	2	6	10	2	1																				
36 Kr	2	2	6	2	6	10	2	6																				
37 Rb	2	2	6	2	6	10	2	6			1																	
38 Sr	2	2	6	2	6	10	2	6			2																	
47 Ag	2	2	6	2	6	10	2	6	10		1																	
48 Cd	2	2	6	2	6	10	2	6	10		2																	
49 In	2	2	6	2	6	10	2	6	10		2	1																
54 X	2	2	6	2	6	10	2	6	10		2	6																
55 Cs	2	2	6	2	6	10	2	6	10		2	6				1												
56 Ba	2	2	6	2	6	10	2	6	10		2	6				2												
79 Au	2	2	6	2	6	10	2	6	10	14	2	6	10			1												
80 Hg	2	2	6	2	6	10	2	6	10	14	2	6	10			2												
81 Tl	2	2	6	2	6	10	2	6	10	14	2	6	10			2	1											
86 Em	2	2	6	2	6	10	2	6	10	14	2	6	10			2	6											
87 --	2	2	6	2	6	10	2	6	10	14	2	6	1			2	6					1						
88 Ra	2	2	6	2	6	10	2	6	10	14	2	6	1[illegible]			2	6					2						
92 U	2	2	6	2	6	10	2	6	10	14	2	6	10	3		2	6	1				2						
100 —	2	2	6	2	6	10	2	6	10	14	2	6	10	11		2	6	1				2						

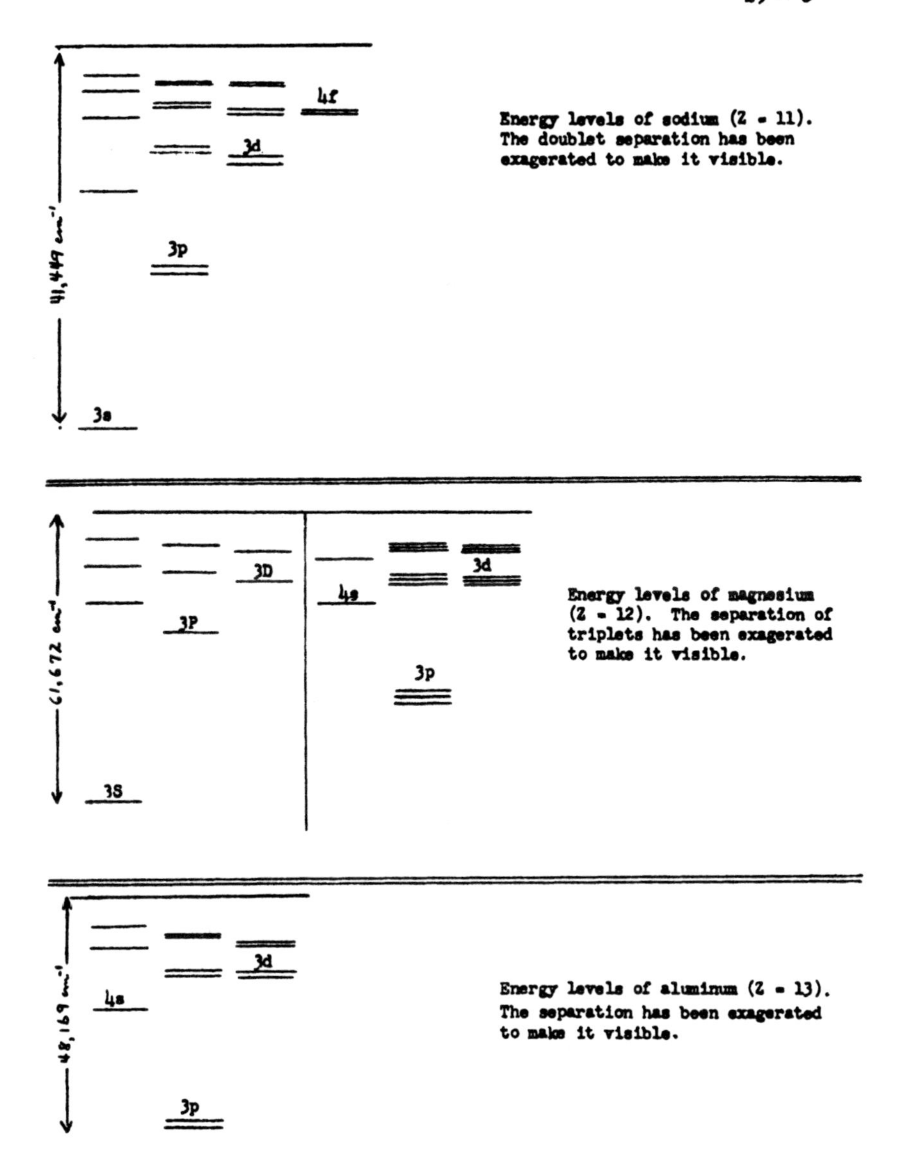

Energy levels of sodium (Z = 11). The doublet separation has been exagerated to make it visible.

Energy levels of magnesium (Z = 12). The separation of triplets has been exagerated to make it visible.

Energy levels of aluminum (Z = 13). The separation has been exagerated to make it visible.

30 - Systems with identical particles

Generalities.

Case of two identical prtcls

$$(1) \quad \begin{aligned} H\,\psi(x_1, x_2) &= E\,\psi(x_1, x_2) \\ H\,\psi(x_2, x_1) &= E\,\psi(x_2, x_1) \end{aligned}$$

Therefore if E non deg. then

$$(2) \quad \psi(x_1 x_2) = k\,\psi(x_2 x_1)$$

but $\psi(x_1 x_2) = k\,\psi(x_2 x_1) = k^2\,\psi(x_1 x_2)$

$$(3) \quad k^2 = 1 \qquad k = \pm 1$$

(4) Either $\psi(x_1, x_2) = \psi(x_2 x_1)$ (symmetric)

or $\psi(x_1, x_2) = -\psi(x_2, x_1)$ (antisymmetric)

If E was deg. two may fail. But instead of base fcts $\psi(x_1 x_2), \psi(x_2 x_1)$ may choose

(5) $\psi(x_1 x_2) + \psi(x_2 x_1)$ (symmetric)

or $\psi(x_1 x_2) - \psi(x_2 x_1)$ (antisym)

Therefore in general.

(6) The e.f.'s of a system with two identical prtcls may always be taken to be either symmetric or antisimmetric

(7) **Theorem**. If $\psi(x_1, x_2, 0)$ is (anti)symmetric, so is $\psi(x_1, x_2, t)$

Because

$$(8) \quad H\begin{Bmatrix}\text{symm function}\\ \text{antisym function}\end{Bmatrix} = \begin{Bmatrix}\text{sym}\\ \text{antisym}\end{Bmatrix}\text{function}$$

Then

$\dot{\psi} = \frac{1}{i\hbar} H\psi$ has same symmetry of ψ.

Then proof by induction from t to $t+dt$

Postulate: Some types of particles (electrons, protons, neutrons, neutrinos, ...) have antisym wave fcts. Others (photons, pions, ...) have symmetric wave functions.

(9)

$$\psi(x_1\, x_2 \dots x_i \dots x_k \dots x_n) = \pm\psi(x_1\, x_2 \dots x_k \dots x_i \dots x_n)$$

+ sign for photons, pions, ...

− sign for electron, protons, neutrons, ...

(10) **Comments**. Pauli has proved that:
antisym particles have half odd spin
simmetric " integral "
No exceptions are "known"

Consider a particle (e.g. an atom) made of other particles (e.g. some electrons, some protons,

(11) some neutrons). For this type of particle parity is $(-1)^N$ where N is the number of antisymmetric

particles entering in its structure.

Examples

H atom, α particle, deuteron } are sym.

Deuterium atom, Tritium nucleus, Nitrogen (N^{14}) atom } are antisym.

Case of independent particles

$$(12) \quad H = H_1 + H_2 + \cdots + H_n$$

H_1 operates on prtcle 1, H_2 " " " 2, - - - e.g. $H_i = \frac{1}{2m_i} p_i^2 + V_i(\vec{x}_i)$

Do not assume at first that ①, ②, ... are identical particles.

Find immediately eigenfunctions

$$(13) \quad \begin{cases} \psi(x_1, x_2 \cdots x_n) = \psi_1(x_1)\, \psi_2(x_2) \cdots \psi_n(x_n) \\ E = E_1 + E_2 + \cdots + E_n \\ \text{where} \quad H_i\, \psi_i(x_i) = E_i\, \psi_i(x_i) \end{cases}$$

Namely: The eigenfunctions of independent particles are products of the eigenfunctions of the individual particles. The corresponding e.v. is the sum of the individual e.v.'s

Assume now particles identical.

Then (13) in general not acceptable because

$$(14) \quad \psi_{n_1}(x_1)\, \psi_{n_2}(x_2) \cdots \psi_{n_m}(x_m)$$

is in general neither sym. nor antisym.

(14) is solution of $H\psi = E\psi$ with

(15) $$E = \sum_{i=1}^{m} E_{n_i}$$

Other deg. solutions with same E are obtained by permuting the lower indices $n_1\, n_2 \cdots n_m$.

Then: form $(n_1, n_2 \cdots n_m) \to (P n_1, P n_2, \cdots, P n_m)$ by permutation P

Symmetric solution

(16) $$\psi_{sym} = \sum_{(P)} \psi_{Pn_1}(x_1)\, \psi_{Pn_2}(x_2) \cdots \psi_{Pn_m}(x_m)$$

For normalized wave function see (21)

$\sum$ over all permutations.

Form antisym solution

(17) $$\psi_{anti} = \sum_{(P)} (-1)^P \psi_{Pn_1}(x_1) \cdots \psi_{Pn_m}(x_m) =$$

or equivalent

(18) $$\psi_{anti} = \begin{vmatrix} \psi_{n_1}(x_1), & \psi_{n_1}(x_2), & \cdots, & \psi_{n_1}(x_m) \\ \psi_{n_2}(x_1), & \psi_{n_2}(x_2), & \cdots, & \psi_{n_2}(x_m) \\ \cdots & \cdots & \cdots & \cdots \\ \psi_{n_m}(x_1), & \psi_{n_m}(x_2), & \cdots, & \psi_{n_m}(x_m) \end{vmatrix}$$

this is a determinant

for normalization factor, see (27)

(16) or (17) will be selected according to the type of particles.

(19) <u>Pauli principle</u>. For <u>antisymmetric</u> particles: Solution (18) obviously vanishes when two or more of individual state indices $n_1, n_2, \cdots, n_m$ are equal. Therefore: For these particles (electrons, protons, neutrons, ...) no state exists in which two identical particles are in the same (completely specified) state.

Occupation numbers.

(20) $N_1\ N_2 \ldots N_s \ldots\quad ,\quad N_1+N_2+\ldots+N_s+\ldots = m$

are no's of id. ptcles in indiv. states $1, 2, \ldots, s, \ldots$

<u>a – Sym. particles</u>: (16) is completely defined by the occupation numbers. Therefore: <u>giving the occ. numbers completely defines the state</u>. Rewrite (16) with normalization factor:

(21) $$\psi_{sim} = \sqrt{\frac{N_1!\,N_2!\ldots N_s!\ldots}{m!}} \sum_{(P)} \psi_{Pn_1}(x_1)\ldots\psi_{Pn_m}(x_m)$$

<u>b – Antisym. particles</u>. Also in this case e.f. (17) or (18) is compl. specified by occ. no's (20). However, only allowable values of occ. no's are 0 and 1. Rewrite (18) with norm. factor

(22) $$\psi_{antis} = \frac{1}{\sqrt{m!}} \begin{vmatrix} \psi_{n_1}(x_1) & \psi_{n_1}(x_2) \cdots & \psi_{n_m}(x_m) \\ \psi_{n_2}(x_1) & \psi_{n_2}(x_2) \cdots & \psi_{n_2}(x_m) \\ \cdots & \cdots & \cdots \\ \psi_{n_m}(x_1) & \psi_{n_m}(x_2) \cdots & \psi_{n_m}(x_m) \end{vmatrix}$$

Discuss here foundation of <u>quantum statistics</u>; Statistical wts of (20):

(23) $$\begin{cases} \text{Boltzmann)}\ \dfrac{N!}{N_1!\,N_2!\ldots}, \quad \text{B.E.)}\ 1\ (\text{one}), \\ \text{F.D.} \begin{cases} 1 & \text{if no occ. no is} > 1 \\ 0 & \text{if some occ. no is} > 1 \end{cases} \end{cases}$$

Discussion & comments: With respect to Boltzmann, B.E. favors bunching, F.D. discourages bunching.

31 - Two electron system.

Notation

(1) $\alpha = \begin{vmatrix} 1 \\ 0 \end{vmatrix} \qquad \beta = \begin{vmatrix} 0 \\ 1 \end{vmatrix}$ α spin up, β spin down

For two electrons, 1 & 2, notation: For example

(2) $\alpha(\zeta_1)\,\beta(\zeta_2) = \alpha\beta$ & similar

Then 4 spin functions:

(3) $\alpha\alpha,\ \alpha\beta,\ \beta\alpha,\ \beta\beta$

Are the base of all two electron spin fcts. Change the base: Total spin

(4) $\vec{S} = \vec{S}_1 + \vec{S}_2$

Make

(5) $\vec{S}^2$ & $\vec{S}_z$ diagonal

Use general method of sect. 28 (or directly)

(6)

Base fcts	$\vec{S}^2$	$\lvert\vec{S}\rvert$	S_z	Spins	Spin symmetry
$\alpha\alpha$	2	1	1	parallel	symm.
$(\alpha\beta+\beta\alpha)/\sqrt{2}$	2	1	0	"	"
$\beta\beta$	2	1	-1	"	"
$(\alpha\beta-\beta\alpha)/\sqrt{2}$	0	0	0	antiparallel	antisym.

(7) Observe: {parallel / antiparallel} spins have spin wave fcts {sym. / antisy.}

Two electron wave fct must be antisymmetric then following possibilities

(8)
$$\alpha\alpha\, u(\vec{x}_1\vec{x}_2),\quad \frac{\alpha\beta+\beta\alpha}{\sqrt{2}}\, u(\vec{x}_1\vec{x}_2),\quad \beta\beta\, u(\vec{x}_1\vec{x}_2)$$
$$\frac{\alpha\beta-\beta\alpha}{\sqrt{2}}\, v(\vec{x}_1\vec{x}_2)$$
with $u(\vec{x}_1\vec{x}_2)$ antisymmetric, $v(\vec{x}_1\vec{x}_2)$ symmetric

Case (a). Two independent electrons

(9) $H_0 = H(1) + H(2)$

Neglect spin orbit interaction

Then let

(10) $H(1)\,\psi_n(\vec{x}_1) = E_n\,\psi_n(\vec{x}_1)$

be soln of one particle problem.

Note: one electron problem has two deg. sol'ns $\alpha\,\psi_n(\vec{x}_1)$, $\beta\,\psi_n(\vec{x}_1)$

Then two electron problem has e.v.'s $E_n + E_m$ with following (degenerate) sol'ns

(11)
$$(\alpha\alpha)\left[\psi_n(x_1)\,\psi_m(x_2) - \psi_m(x_1)\,\psi_n(x_2)\right]/\sqrt{2}$$
or
$$\frac{\alpha\beta+\beta\alpha}{\sqrt{2}}\left[\quad \text{same} \quad\right]/\sqrt{2}$$
or
$$(\beta\beta)\left[\quad \text{same} \quad\right]/\sqrt{2}$$

These have $S=1$ orbital antisym. spin symmetric

or
$$\frac{\alpha\beta-\beta\alpha}{\sqrt{2}}\left[\psi_n(x_1)\,\psi_m(x_2) + \psi_m(x_1)\,\psi_n(x_2)\right]/\sqrt{2}$$

This has $S=1$ orbital sym. spin antisym.

Introduce now Coulomb interaction

(12)
$$H_{coulomb} = \frac{e^2}{|\vec{x}_1 - \vec{x}_2|} = \frac{e^2}{r_{12}}$$

Treat (12) as perturbation (first order)

(13)
$$\delta E_{coul} = \overline{H_{coul}} = \iint \sum_{spin} d^3x_1\, d^3x_2 \left|\text{wave fct}\right|^2.$$

Result different for $S=1$ (triplet) states and $S=0$ (singlet) states (comment: no off diagonal elements). Find

~~δE(triplet)~~ (struck out)

(upper sign for triplets, lower " " singlets)

$$(14)\quad \delta E_{coulomb} = \iint \frac{e^2}{r_{12}} |\psi_1(x_1)|^2 |\psi_2(x_2)|^2 d\vec{x}_1 d\vec{x}_2 \mp$$

(This is electrostatic ~~unt~~ ~~ad~~ energy)

(assume ψ_1; ψ_2 real)

$$\mp \iint \frac{e^2}{r_{12}} \psi_1(x_1)\psi_2(x_1)\psi_1(x_2)\psi_2(x_2) d\vec{x}_1 d\vec{x}_2$$

This is exchange integral

Discussion & comments on this formula
Exchange integral as an apparent very strong spin spin coupling.
Relationship to theory of ferromagnetism.
Role of spin orbit interactions and triplet splitting.

The He spectrum.

Parahelium (Singlet)	$1s^2\ {}^1S_0 = 198305$	$2p1s\ {}^1P_0 = 27176$
	$2s1s\ {}^1S_0 = 32033$	$3d1s\ {}^1D_0 = 12206$
	$3s1s\ {}^1S_0 = 19446$	
Orthohelium (Triplet)	$2s1s\ {}^3S_1 = 38455$	$2p1s\ {}^3P_0 = 29223.87$
		" $\ {}^3P_1 = 29223.799$
	$3s1s\ {}^3S_1 = 15074$	" $\ {}^3P_2 = 29222.878$

(Comments)

Ritz with trial fct $e^{-\alpha\frac{r_1+r_2}{a}}$ gives $\alpha = \frac{27}{16}$
Ground level $(2\times\frac{27^2}{16^2} - 4)$ Rydberg $= 186{,}000\ \text{cm}^{-1}$

32 – Hydrogen molecule

Generalities on molecular spectra

Rotational oscillation and electronic levels.

Electronic levels of H_2 – molecule

①

②

a r b

(1) $$H = \frac{p_1^2 + p_2^2}{2m} + \frac{e^2}{r} + \frac{e^2}{r_{12}} - \frac{e^2}{r_{a1}} - \frac{e^2}{r_{a2}} - \frac{e^2}{r_{b1}} - \frac{e^2}{r_{b2}}$$

Heitler London method.

Discuss two zero approx wave fcts

(2) $\psi = a(1)\, b(2) \pm a(2)\, b(1)$ + for S=0 (singlet) − for S=1 (triplet)

$a(1)$, $b(1)$ are hydrogen wave fcts for electron ① near nucleus $\underline{a}$ or $\underline{b}$.

Step ⓐ: normalization

(3) $$\int \psi^2 d\vec{x}_1\, d\vec{x}_2 = \left(\int a^2(1)\, dx_1\right)\left(\int b^2(2)\, dx_2\right) + \left(\int a^2(2)\, dx_2\right)\left(\int b^2(1)\, dx_1\right)$$
$$\pm 2 \int a(1)\, b(1)\, dx_1 \int a(2)\, b(2)\, dx_2$$
$$= 2(1 + \beta^2)$$

(4) $$\beta = \int a(1)\, b(1)\, d\vec{x}_1$$

Normalized wave fcts

(5) $$\psi_\pm = \frac{a(1)\, b(2) \pm a(2)\, b(1)}{\sqrt{2(1 \pm \beta^2)}}$$

(6) $$E_\pm = \iint \psi_\pm H \psi_\pm \, dx_1 \, dx_2$$

Use

(7) $$\left(\frac{1}{2m} p_1^2 - \frac{e^2}{r_{a1}}\right) a(1) = - R\, a(1)$$

R = Rydberg energy = +13.6 eV

Find

(8) $$H a(1) b(2) = \left(-2R + \frac{e^2}{r} + \frac{e^2}{r_{12}} - \frac{e^2}{r_{a2}} - \frac{e^2}{r_{b1}}\right) a(1) b(2)$$

Find

(9) $$E_\pm = -2R + \frac{e^2}{r} + \frac{1}{1 \pm \beta^2} \iint \left(\frac{e^2}{r_{12}} - \frac{e^2}{r_{a2}} - \frac{e^2}{r_{b1}}\right) a^2(1)\, b^2(2)\, dx_1 dx_2$$
$$\pm \frac{1}{1 \pm \beta^2} \iint \left(\frac{e^2}{r_{12}} - \frac{e^2}{r_{a2}} - \frac{e^2}{r_{b1}}\right) a(1) b(1) a(2) b(2)\, dx_1 dx_2$$

<u>Discussion</u>

Take $-2R$ as zero energy (energy of two distant atoms)

Term $\frac{e^2}{r}$ is potential energy of nuclei

first $\iint$-term (apart of small β) is mutual electrostatic interaction of two electron clouds $e\,a^2(1)$ and $e\,b^2(2)$ between each other and with the other nucleus.

Second $\iint$ is exchange integral. This is negative and depends on r as follows

Adding various term find

No binding for E_-

Binding for E_+

For ground state of H_2 two electrons have then opposite spins $(S=0)$

E_-, E_+, r, equil distance

Heitler London method sketched above is quantitatively poor.

Better for ground state <u>Wang</u> method with Ritz trial fct

$$\psi(x_1, x_2) = e^{-\frac{z}{a}(r_{a1} + r_{b2})} + e^{-\frac{z}{a}(r_{b1} + r_{a2})} \tag{10}$$

a = Bohr radius

z = adjustable parameter of Ritz method

Minimize for each value of r

$$\bar{H} = \frac{\int \psi(x_1 x_2)\, H\, \psi(x_1 x_2)\, d\vec{x}_1 d\vec{x}_2}{\int |\psi(x_1 x_2)|^2 d\vec{x}_1 d\vec{x}_2} \tag{11}$$

(12)

	Wang	Experiment
Bind. Energy	.278 Rydberg	.326 Rydberg
Mom. of inertia	$.459 \times 10^{-40}$	$.467 \times 10^{-40}$
Oscill. frequency	4900 cm^{-1}	4360 cm^{-1}

Rotational levels (Role of nuclear spin)
Approx. hamiltonian for rotational levels only

~~(see Sect 2~~

(13) $-\frac{\hbar^2}{2A}\Lambda$ [See Sect 2 (14)]

A = mom. of inertia

yields rot. energy levels

(14) $\begin{cases} \frac{\hbar^2}{2A} l(l+1) & l = 0, 1, 2, \dots \\ \psi_l = Y_{lm}(\vartheta, \varphi) \end{cases}$

(14) applies to diatomic molecules when there is no resultant ang. mom. of the electrons along figure axis.

Even in this case, however, complications for identical nuclei.

Example: two nuclei identical with nuclear spin 0, and B.E. statistics require symmetric wave fnct. Now $Y_{lm}(\vartheta, \varphi)$ sym for interchange of nuclei only when l even. Therefore in this case all odd $\underline{l}$'s are absent

(Comment as to possible complications due to symmetry of electronic levels)

For hydrogen, the two protons have spin 1/2 and antisym. wave fcts

Therefore (like for two electron system) rotational terms split into

Para hydrogen terms

Nuclear spins antiparallel $\ell = 0, 2, 4, \ldots$

and

Orthohydrogen terms

Nuclear spins parallel $\ell = 1, 3, 5, \ldots$

Comments. Alternating band intensities

Very slow ortho-para conversion in hydrogen

Specific heat of hydrogen rotation.

Topics for discussion — Band spectra of diatomic molecules.

33 – Collision theory

Scattering by short range central force field.

(1) $\psi \to e^{ikz} - f(\theta)\frac{e^{ikr}}{r}$ (asymptotic for $r\to\infty$)

(2) $k = \frac{1}{\hbar}p$

(1) yields diff cross sect

(3) $\frac{d\sigma}{d\omega} = |f(\theta)|^2$

Develop (1) in sph. harmonics by

(4) $e^{ikz} = \frac{\pi\sqrt{2}}{\sqrt{kr}}\sum_{l=0}^{\infty} i^l\sqrt{2l+1}\,Y_{l,0}(\theta)\,J_{l+\frac{1}{2}}(kr)$

Also use

$$J_n(x) \longrightarrow \sqrt{\frac{2}{\pi x}}\cos\left(x - \frac{\pi}{4} - \frac{\pi n}{2}\right)$$

(5) $e^{ikz} \to \frac{\sqrt{4\pi}}{kr}\sum_0^\infty i^l\sqrt{2l+1}\,Y_{l0}\sin\left(kr - \frac{\pi l}{2}\right) =$

$$= \frac{\sin kr}{kr} + \ldots$$

Also dev. $f(\theta)$ in sph. harm. by

(6) $f(\theta) = \sum_l a_l P_l(\cos\vartheta) = \sqrt{4\pi}\sum_l \frac{a_l}{\sqrt{2l+1}}Y_{l0}(\theta)$

(7) $\psi \longrightarrow \frac{\sqrt{4\pi}}{kr}\sum_l \frac{Y_{l0}}{\sqrt{2l+1}}\left\{e^{ikr}\left[-a_l - \frac{i}{2}\frac{2l+1}{k}\right] + \right.$

$$\left. + e^{-ikr}(-1)^l\frac{i}{2}\frac{2l+1}{k}\right\}$$

Comments — In- and outgoing wave ~~have~~ must have = amplitudes. Then

(8) $$a_l + \frac{i}{2}\frac{2l+1}{k} = e^{2i\alpha_l}\left(\frac{i}{2}\frac{2l+1}{k}\right)$$

or

(9) $$a_l = \frac{i}{2}\frac{2l+1}{k}\left(e^{2i\alpha_l}-1\right)$$

and radial wave fct of l, $R_l(r) = \frac{u_l(r)}{r}$

(10) $$u_l(r) \xrightarrow{r} \sin\left(kr - \frac{\pi l}{2} + \alpha_l\right)$$

← phase shift.

Determine α_l from radial equation

(11) $$\begin{cases} u_l''(r) - \dfrac{l(l+1)}{r^2}u_l + \dfrac{2m}{\hbar^2}\left[E - U(r)\right]u_l = 0 \\ E = \dfrac{\hbar^2}{2m}k^2 \end{cases}$$

(12) $$u_l'' + \left\{k^2 - \frac{2m}{\hbar^2}U(r) - \frac{l(l+1)}{r^2}\right\}u_l = 0$$

Solution behavior ~~for~~ r small & large

(13) $$r^{l+1} \longleftarrow u_l(r) \longrightarrow \text{const} \times \sin\left(kr + \alpha_l - \frac{\pi l}{2}\right)$$

determines α_l.

Express $\frac{d\sigma}{d\omega}$ in terms of α_l (use (9), (6), (3))

(14) $$\frac{d\sigma}{d\omega} = \frac{1}{4k^2}\left|\sum_l (2l+1)P_l(\cos\vartheta)\left(e^{2i\alpha_l}-1\right)\right|^2$$

Integrate:

(15) $\sigma = 4\pi \lambda\!\!\!^{-2} \sum_{\ell} (2\ell+1) \sin^2 \alpha_\ell$ $\quad \lambda\!\!\!^{-} = 1/k$

α_0 at low energy only important $\ell = 0$

(16) $\left\{ \begin{array}{l} \alpha_0 = -k \times \text{scattering length} = -kb_0 \\ \text{(at low energy)} \end{array} \right.$

Then at low energy

(17) $\sigma \to 4\pi b^2$

One can proove that in simple cases at low energy

$$\alpha_\ell \sim k^{2\ell+1}$$

Comments — Examples — Coulomb forces (See Schiff Sect. 20) — Scattering by hard sphere Absorption & shadow scattering —

34 – Dirac's theory of the free electron

Time dep. Schrödinger eq. for particle

$$i\hbar\frac{\partial\psi}{\partial t}=-\frac{\hbar^2}{2m}\left(\frac{\partial^2\psi}{\partial x^2}+\frac{\partial^2\psi}{\partial y^2}+\frac{\partial^2\psi}{\partial z^2}\right)$$

treats t, x, y, z very non symmetrically. Search for relativistic equation for electron of first order in t, x, y, z.

Notation

(1)
$$x=x_1 \quad y=x_2 \quad z=x_3 \quad ict=x_4 \quad (ct=x_0)$$
$$p_x=\frac{\hbar}{i}\frac{\partial}{\partial x} \text{ or } p_i=\frac{\hbar}{i}\frac{\partial}{\partial x_i}$$
$$p_4=\frac{\hbar}{i}\frac{\partial}{\partial x_4}=-\frac{\hbar}{c}\frac{\partial}{\partial t}=\frac{i}{c}E$$

(use $E=i\hbar\frac{\partial}{\partial t}$)

(2) Ordinary vectors
$$\vec{x}\equiv(x_1,x_2,x_3) \qquad \vec{p}=(p_1,p_2,p_3)$$

(3) Four vectors
$$\underset{\sim}{x}\equiv(x_1\, x_2\, x_3\, x_4) \text{ or } \underset{\sim}{p}\equiv(p_1\, p_2\, p_3\, p_4)$$

If ψ were a scalar, simplest first order eqn would be (constant coeff.)

$$\psi=a^{(1)}\frac{\partial\psi}{\partial x_1}+a^{(2)}\frac{\partial\psi}{\partial x_2}+a^{(3)}\frac{\partial\psi}{\partial x_3}+a^{(4)}\frac{\partial\psi}{\partial x_4}=\frac{i}{\hbar}a^{(\mu)}p_\mu\psi$$

(Sum over equal indices)

It will prove necessary however to take ψ to have several (four) components. Instead of above, write

(4)
$$imc\,\psi_k=\gamma^{(\mu)}_{kl}\,p_\mu\,\psi_l=\frac{\hbar}{i}\gamma^{(\mu)}_{kl}\frac{\partial\psi_l}{\partial x_\mu}$$

In matrix notation: ψ a vertical slot of (four) elements $\gamma_\mu = \| \gamma^{(\mu)}_{kl} \|$ a square matrix (four × four matrix)

(5) $\quad i m c \psi = \gamma_\mu p_\mu \psi \quad$ (sum over μ)

$$= \frac{\hbar}{i} \gamma_\mu \frac{\partial \psi}{\partial x_\mu}$$

$p_\mu = \frac{\hbar}{i} \frac{\partial}{\partial x_\mu}$ operates on dependence of ψ on $\underset{\sim}{x}$

γ_μ operates on an internal variable similar to the spin variable of Pauli, however with 4 components as will be seen. Follows:

(6) $\{ \quad \gamma_\mu$ commutes with p_ν and x_ν

From (5)

$$(imc)^2 \psi = (\gamma_\mu p_\mu)^2 \psi$$

Or (omitting ψ) $\qquad$ use (1) $\quad p_4^2 = -\frac{E^2}{c^2}$

use (6)

$$-m^2c^2 = \gamma_1^2 p_1^2 + \gamma_2^2 p_2^2 + \gamma_3^2 p_3^2 - \gamma_4^2 \frac{E^2}{c^2} +$$

$$+ (\gamma_1\gamma_2 + \gamma_2\gamma_1) p_1 p_2 + \text{similar terms}$$

This can be identified with the relativistic momentum energy relation

(7) $\quad m^2c^2 + \vec{p}^{\,2} = \frac{E^2}{c^2} \qquad$ by postulating

(8) $\quad \gamma_1^2 = \gamma_2^2 = \gamma_3^2 = \gamma_4^2 = 1 \qquad \gamma_\mu\gamma_\nu + \gamma_\nu\gamma_\mu = 0$ for $\mu \neq \nu$

One finds that the lowest order matrices for which (8) can be fulfilled is the 4-th. For order four there are many solutions that are essentially equivalent. We choose the "standard" solution

$$(9)\quad \gamma_1 = \begin{vmatrix} 0 & 0 & 0 & -i \\ 0 & 0 & -i & 0 \\ 0 & i & 0 & 0 \\ i & 0 & 0 & 0 \end{vmatrix}; \quad \gamma_2 = \begin{vmatrix} 0 & 0 & 0 & -1 \\ 0 & 0 & 1 & 0 \\ 0 & 1 & 0 & 0 \\ -1 & 0 & 0 & 0 \end{vmatrix}; \quad \gamma_3 = \begin{vmatrix} 0 & 0 & -i & 0 \\ 0 & 0 & 0 & i \\ i & 0 & 0 & 0 \\ 0 & -i & 0 & 0 \end{vmatrix}$$

and

$$(10)\quad \beta = \gamma_4 = \begin{vmatrix} 1 & 0 & 0 & 0 \\ 0 & 1 & 0 & 0 \\ 0 & 0 & -1 & 0 \\ 0 & 0 & 0 & -1 \end{vmatrix}$$

$\gamma_1, \gamma_2, \gamma_3$ act in many ways as the components of a vector and will be denoted by

$$(11)\quad \vec{\gamma} = (\gamma_1, \gamma_2, \gamma_3) \quad \text{also} \quad \underset{\sim}{\gamma} \equiv (\gamma_1\ \gamma_2\ \gamma_3\ \gamma_4)$$

four vector

Then (5) becomes

$$(12)\quad imc\,\psi = \left(\vec{\gamma}\cdot\vec{p} + \frac{i}{c}E\gamma_4\right)\psi = \underset{\sim}{\gamma}\cdot\underset{\sim}{p}\,\psi$$

Multiply to left by $\gamma_4 = \beta$ using $\gamma_4^2 = \beta^2 = 1$

$$(13)\quad \boxed{E\psi = \left(mc^2\beta + c\,\vec{\alpha}\cdot\vec{p}\right)\psi}$$

where

$$(14)\quad \vec{\alpha} = i\beta\vec{\gamma} \quad (\text{or } \alpha_1 = i\beta\gamma_1 \quad \alpha_2 = i\beta\gamma_2 \quad \alpha_3 = i\beta\gamma_3)$$

$$(15)\quad \alpha_1 = \begin{vmatrix} 0 & 0 & 0 & 1 \\ 0 & 0 & 1 & 0 \\ 0 & 1 & 0 & 0 \\ 1 & 0 & 0 & 0 \end{vmatrix}; \quad \alpha_2 = \begin{vmatrix} 0 & 0 & 0 & -i \\ 0 & 0 & i & 0 \\ 0 & -i & 0 & 0 \\ i & 0 & 0 & 0 \end{vmatrix} \quad \alpha_3 = \begin{vmatrix} 0 & 0 & 1 & 0 \\ 0 & 0 & 0 & -1 \\ 1 & 0 & 0 & 0 \\ 0 & -1 & 0 & 0 \end{vmatrix}$$

Properties (check directly)

$$(16) \quad \beta^2 = \alpha_1^2 = \alpha_2^2 = \alpha_3^2 = 1$$

$$(17) \begin{cases} \beta\alpha_1 + \alpha_1\beta = 0 \quad \beta\alpha_2 + \alpha_2\beta = 0 \quad \beta\alpha_3 + \alpha_3\beta = 0 \\ \alpha_1\alpha_2 + \alpha_2\alpha_1 = 0 \quad \alpha_2\alpha_3 + \alpha_3\alpha_2 = 0 \quad \alpha_3\alpha_1 + \alpha_1\alpha_3 = 0 \end{cases}$$

(18) { β & the α's have square = unit matrix
β & the α's anticommute with each other.
β & the α's are hermitian

One can prove that all the physical consequences of (13) do not depend on the special choice (10), (15) of $\alpha_1\ \alpha_2\ \alpha_3\ \beta$. They would be the same if a different set of four 4×4 matrices with the specifications (18) had been chosen. In particular its is possible by unitary transformation to interchange the roles of the four matrices. So that their differences are only apparent.

(19) { Check that for each of the 7 matrices $\gamma_4 = \beta, \alpha_1, \alpha_2, \alpha_3, \gamma_1, \gamma_2, \gamma_3$ the eigenvalues are $+1$, twice and -1 twice

(13) is written also

(20) $$E\psi = H\psi$$

(21) $$\begin{cases} H = \text{hamiltonian} \\ H = mc^2\beta + c\,\vec{\alpha}\cdot\vec{p} \end{cases}$$ for $\psi = \begin{vmatrix} \psi_1 \\ \psi_2 \\ \psi_3 \\ \psi_4 \end{vmatrix}$

Time indep. equation

(22) $$\begin{cases} E\psi_1 = mc^2\psi_1 + \frac{c\hbar}{i}\left\{\frac{\partial\psi_4}{\partial x} - i\frac{\partial\psi_4}{\partial y} + \frac{\partial\psi_3}{\partial z}\right\} \\ E\psi_2 = mc^2\psi_2 + \frac{c\hbar}{i}\left\{\frac{\partial\psi_3}{\partial x} + i\frac{\partial\psi_3}{\partial y} - \frac{\partial\psi_4}{\partial z}\right\} \\ E\psi_3 = -mc^2\psi_3 + \frac{c\hbar}{i}\left\{\frac{\partial\psi_2}{\partial x} - i\frac{\partial\psi_2}{\partial y} + \frac{\partial\psi_1}{\partial z}\right\} \\ E\psi_4 = -mc^2\psi_4 + \frac{c\hbar}{i}\left\{\frac{\partial\psi_1}{\partial x} + i\frac{\partial\psi_1}{\partial y} - \frac{\partial\psi_2}{\partial z}\right\} \end{cases}$$

also time dep. Schr. eq by $E \to i\hbar\frac{\partial}{\partial t}$

Plane wave solution. Take

(23) $$\psi = \begin{vmatrix} u_1 \\ u_2 \\ u_3 \\ u_4 \end{vmatrix} e^{\frac{i}{\hbar}\vec{p}\cdot\vec{x}}$$ $\vec{p}$ now a numerical vector

$u_1\ u_2\ u_3\ u_4$ are constants.

Substitute in (22) (Divide by common exp. factor)

(24) $$\begin{cases} Eu_1 = mc^2u_1 + c(p_x - ip_y)u_4 + cp_z u_3 \\ Eu_2 = mc^2u_2 + c(p_x + ip_y)u_3 - cp_z u_4 \\ Eu_3 = -mc^2u_3 + c(p_x - ip_y)u_2 + cp_z u_1 \\ Eu_4 = -mc^2u_4 + c(p_x + ip_y)u_1 - cp_z u_2 \end{cases}$$

Four homog. linear eq. for $u_1\ u_2\ u_3\ u_4$.
Require det = 0. One finds e.v's of E

(25) $E = +\sqrt{m^2c^4 + c^2p^2}$ twice and $E = -\sqrt{m^2c^4 + c^2p^2}$ twice)

For each $\vec{p}$, E has twice the value $E = \sqrt{m^2c^4 + c^2p^2}$ but also twice the negative value $E = -\sqrt{m^2c^4+c^2p^2}$ (Comments)

A set of 4 orthogonal normalized spinors u is

(26) For $E = +\sqrt{m^2c^4+c^2p^2} = R$

$$u^{(1)} = \sqrt{\frac{mc^2+R}{2R}} \begin{vmatrix} 1 \\ 0 \\ \frac{cp_z}{mc^2+R} \\ \frac{c(p_x+ip_y)}{mc^2+R} \end{vmatrix} \quad \text{or} \quad u^{(2)} = \sqrt{\frac{mc^2+R}{2R}} \begin{vmatrix} 0 \\ 1 \\ \frac{c(p_x-ip_y)}{mc^2+R} \\ \frac{-cp_z}{mc^2+R} \end{vmatrix}$$

(27) For $E = -R = -\sqrt{m^2c^4+c^2p^2}$

$$u^{(3)} = \sqrt{\frac{R-mc^2}{2R}} \begin{vmatrix} \frac{cp_z}{R-mc^2} \\ \frac{c(p_x+ip_y)}{R-mc^2} \\ 1 \\ 0 \end{vmatrix} \quad \text{or} \quad u^{(4)} = \sqrt{\frac{R-mc^2}{2R}} \begin{vmatrix} \frac{c(p_x-ip_y)}{R-mc^2} \\ \frac{-cp_z}{R-mc^2} \\ 0 \\ 1 \end{vmatrix}$$

Observe: for $|p| < mc$ the third + fourth component of the positive energy solutions $u^{(1)}$ + $u^{(2)}$ are very small and the first and second component of the neg. en. solutions $u^{(3)}$ + $u^{(4)}$ are very small (of order p/mc).

Meaning of neg. + pos. energy levels.

The Dirac sea — Vacuum state

Positrons as holes.

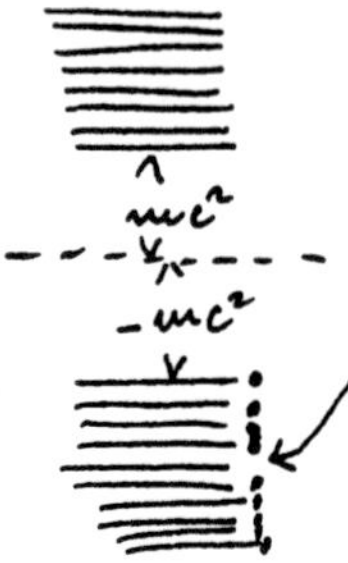

Mom + energy of the positron are $(-\vec{p} + -\vec{E})$ of the "hole" state.

(28) $u^{(1)} e^{\frac{i}{\hbar}\vec{p}\cdot\vec{x}}, \; u^{(2)} e^{\frac{i}{\hbar}\vec{p}\cdot\vec{x}}$ electron states (spin up + down) (mom. $= \vec{p}$, energy $= +\sqrt{m^2c^4+c^2p^2}$)

(29) $\left.\begin{array}{l} u^{(3)} e^{\frac{i}{\hbar}\vec{p}\cdot\vec{x}} \\ u^{(4)} e^{\frac{i}{\hbar}\vec{p}\cdot\vec{x}} \end{array}\right\}$ are positron states with momentum $= -\vec{p}$, energy $= +\sqrt{m^2c^4+c^2p^2}$

Given $\psi = u\, e^{\frac{i}{\hbar}p\cdot x}$ (u = 4 component spinor)

it is important to have two operators $\mathcal{P}$ & $\mathcal{N}$ (<u>projection operators</u>) such that $\mathcal{P}\psi$ contains only electron wave fcts, $\mathcal{N}\psi$ contains only neg. energy wave fcts (positron states). $\mathcal{P}$, $\mathcal{N}$

(30) are spinor operators ~~such~~ defined by $\mathcal{P}u^{(1)} = u^{(1)}$, $\mathcal{P}u^{(2)} = u^{(2)}$, $\mathcal{P}u^{(3)} = 0$, $\mathcal{P}u^{(4)} = 0$ and

(31) $\mathcal{N}u^{(1)} = 0, \quad \mathcal{N}u^{(2)} = 0, \quad \mathcal{N}u^{(3)} = u^{(3)}, \quad \mathcal{N}u^{(4)} = u^{(4)}$

These properties define uniquely $\mathcal{P}$ & $\mathcal{N}$

Observe: $Hu^{(1)} = Ru^{(1)}$, $Hu^{(2)} = Ru^{(2)}$, $Hu^{(3)} = -Ru^{(3)}$
$Hu^{(4)} = -Ru^{(4)}$
with
$$R = +\sqrt{m^2c^4 + c^2p^2} \quad (\vec{p} \text{ here a c-vector})$$
and H from (21). Then

$$(32) \quad \mathcal{P} = \frac{1}{2} + \frac{1}{2R} H \quad ; \quad \mathcal{N} = \frac{1}{2} - \frac{1}{2R} H$$

Angular momentum. From (21)

$$(33) \quad [H, xp_y - yp_x] = \frac{\hbar c}{i}(\alpha_1 p_y - \alpha_2 p_x) \neq 0$$

Therefore $xp_y - yp_x$ not a time constant for free Dirac electron. However

$$(34) \quad xp_y - yp_x + \frac{1}{2}\frac{\hbar}{i}\alpha_1\alpha_2 = \hbar J_z$$

Commutes with H. Interpret $\hbar J_z$ as z component of ang. mom.

$$(35) \quad \hbar \vec{J} = \underbrace{\vec{x} \times \vec{p}}_{\text{orbital part}} + \underbrace{\frac{\hbar}{2i}\begin{cases}\alpha_2\alpha_3 \\ \alpha_3\alpha_1 \\ \alpha_1\alpha_2\end{cases}}_{\text{spin part}} = \vec{x} \times \vec{p} + \frac{\hbar}{2}\vec{\sigma}'$$

with

$$(36) \quad \sigma'_x = \frac{1}{i}\alpha_2\alpha_3 = \begin{vmatrix} 0 & 1 & 0 & 0 \\ 1 & 0 & 0 & 0 \\ 0 & 0 & 0 & 1 \\ 0 & 0 & 1 & 0 \end{vmatrix}; \quad \sigma'_y = \frac{1}{i}\alpha_3\alpha_1 = \begin{vmatrix} 0 & -i & 0 & 0 \\ i & 0 & 0 & 0 \\ 0 & 0 & 0 & -i \\ 0 & 0 & i & 0 \end{vmatrix}; \quad \sigma'_z = \frac{1}{i}\alpha_1\alpha_2 = \begin{vmatrix} 1 & 0 & 0 & 0 \\ 0 & -1 & 0 & 0 \\ 0 & 0 & 1 & 0 \\ 0 & 0 & 0 & -1 \end{vmatrix}$$

(Observe analogy with Pauli operators $\vec{\sigma}$ + $\vec{\sigma}'$)

35 – Dirac electron in electromagnetic field

Notation

$$(1)\quad \begin{cases} \vec{A} = (A_1, A_2, A_3) = \text{vector potential} \\ A_4 = i\varphi = (i \times \text{scalar potential}) \\ \underset{\sim}{A} \equiv (A_1, A_2, A_3, A_4) = \text{4-vector potential} \end{cases}$$

$$(2)\quad F_{ik} = \frac{\partial A_k}{\partial x_i} - \frac{\partial A_i}{\partial x_k} = \text{antisym. tensor "electromagnetic field"}$$

$$(3)\quad \begin{cases} (F_{12}, F_{23}, F_{31}) \equiv \vec{B} = \text{magnetic field} \\ (F_{41}, F_{42}, F_{43}) \equiv i\vec{E} \quad (\vec{E} = \text{electric field}) \end{cases}$$

Introduce e.m. field in Dirac equation (34-(12) or (20)(21)) by

$$(4)\quad \vec{p} \rightarrow \vec{p} - \frac{e}{c}\vec{A} \qquad E \rightarrow E - e\varphi$$

or equivalents

$$(5)\quad \begin{cases} \underset{\sim}{p} \rightarrow \underset{\sim}{p} - \frac{e}{c}\underset{\sim}{A} \\ \frac{\partial}{\partial x_\ell} \rightarrow \frac{\partial}{\partial x_\ell} - \frac{ie}{\hbar c}A_\ell \qquad (\ell = 1, 2, 3, 4) \\ \underset{\sim}{\nabla} \rightarrow \underset{\sim}{\nabla} - \frac{ie}{\hbar c}\underset{\sim}{A} \end{cases}$$

Find equivalent equations

$$(6)\quad imc\,\psi = \underset{\sim}{\gamma} \cdot \left(\underset{\sim}{p} - \frac{e}{c}\underset{\sim}{A}\right)\psi$$

or

$$(7) \qquad \left(\frac{mc}{\hbar} + \underset{\sim}{\gamma} \cdot \underset{\sim}{\nabla} - \frac{ie}{\hbar c} \underset{\sim}{A} \cdot \underset{\sim}{\gamma}\right)\psi = 0$$

or

$$(8) \qquad E\psi = H\psi$$

with hamiltonian

$$(9) \qquad H = +e\varphi - e\vec{A}\cdot\vec{\alpha} + mc^2\beta + c\vec{\alpha}\cdot\vec{p}$$

(8) is equiv to four eq.ns similar to (34-(22))

$$(10)\begin{cases}(E - e\varphi - mc^2)\psi_1 = \frac{c\hbar}{i}\left(\frac{\partial\psi_4}{\partial x} - i\frac{\partial\psi_4}{\partial y} + \frac{\partial\psi_3}{\partial z}\right) - e\{(A_x - iA_y)\psi_4 + A_z\psi_3\} \\ (E - e\varphi - mc^2)\psi_2 = \frac{c\hbar}{i}\left(\frac{\partial\psi_3}{\partial x} + i\frac{\partial\psi_3}{\partial y} - \frac{\partial\psi_4}{\partial z}\right) - e\{(A_x + iA_y)\psi_3 - A_z\psi_4\} \\ (E - e\varphi + mc^2)\psi_3 = \frac{c\hbar}{i}\left(\frac{\partial\psi_2}{\partial x} - i\frac{\partial\psi_2}{\partial y} + \frac{\partial\psi_1}{\partial z}\right) - e\{(A_x - iA_y)\psi_2 + A_z\psi_1\} \\ (E - e\varphi + mc^2)\psi_4 = \frac{c\hbar}{i}\left(\frac{\partial\psi_1}{\partial x} + i\frac{\partial\psi_1}{\partial y} - \frac{\partial\psi_2}{\partial z}\right) - e\{(A_x + iA_y)\psi_1 - A_z\psi_2\}\end{cases}$$

Introduce two dicotomic variables

$$(11) \qquad u = \begin{vmatrix}\psi_1 \\ \psi_2\end{vmatrix} \qquad v = \begin{vmatrix}\psi_3 \\ \psi_4\end{vmatrix}$$

And the Pauli operators $\vec{\sigma} = (\sigma_x, \sigma_y, \sigma_y)$. (10) becomes

$$(12)\begin{cases}\frac{i}{c\hbar}(E - mc^2 - e\varphi)\,u = \vec{\sigma}\cdot\left(\vec{\nabla} - \frac{ie}{c\hbar}\vec{A}\right)v \\ \frac{i}{c\hbar}(E + mc^2 - e\varphi)\,v = \vec{\sigma}\cdot\left(\vec{\nabla} - \frac{ie}{c\hbar}\vec{A}\right)u\end{cases}$$

$$(13)\begin{cases}\frac{1}{c}(E - mc^2 - e\varphi)\,u = \vec{\sigma}\cdot\left(\vec{p} - \frac{e}{c}\vec{A}\right)v \\ \frac{1}{c}(E + mc^2 - e\varphi)\,v = \vec{\sigma}\cdot\left(\vec{p} - \frac{e}{c}\vec{A}\right)u\end{cases}$$

Eliminate $\underline{v}$ from (13):

$$\frac{1}{c^2}(E+mc^2-e\varphi)(E-mc^2-e\varphi)u = \frac{1}{c^2}\left\{(E-e\varphi)^2-m^2c^4\right\}u =$$

$$= \frac{1}{c}(E+mc^2-e\varphi)\,\vec{\sigma}\cdot\left(\vec{p}-\frac{e}{c}\vec{A}\right)v =$$

$$=\left\{\left(\vec{\sigma}\cdot\vec{p}-\frac{e}{c}\vec{A}\right)\frac{E+mc^2-e\varphi}{c} - \frac{e}{c^2}\vec{\sigma}\cdot[E,\vec{A}] - \frac{e}{c}\vec{\sigma}\cdot[\varphi,\vec{p}]\right\}v$$

$$=\left(\vec{\sigma}\cdot\vec{p}-\frac{e}{c}\vec{A}\right)^2 u + \left(\frac{e\hbar}{ic^2}\vec{\sigma}\cdot\frac{\partial\vec{A}}{\partial t} + \frac{e\hbar}{ic}\vec{\sigma}\cdot\vec{\nabla}\varphi\right)v =$$

$$=\left(\vec{p}-\frac{e}{c}\vec{A}\right)^2 u + i\vec{\sigma}\cdot\underbrace{\left(\vec{p}-\frac{e}{c}\vec{A}\right)\times\left(\vec{p}-\frac{e}{c}\vec{A}\right)}\,u - \frac{e\hbar}{ic}(\vec{\sigma}\cdot\vec{\mathcal{E}})v$$

$$-\frac{e}{c}(\vec{p}\times\vec{A}+\vec{A}\times\vec{p})$$

$$\|$$

$$-\frac{e}{c}\frac{\hbar}{i}\nabla\times A = -\frac{e\hbar}{ci}B$$

($\vec{\mathcal{E}}$ = electric field $= -\nabla\varphi - \frac{1}{c}\frac{\partial A}{\partial t}$)

Find then

$$(14)\quad \left\{\frac{(E-e\varphi)^2}{c^2} - m^2c^2 - \left(\vec{p}-\frac{e}{c}\vec{A}\right)^2\right\}u = -\frac{e\hbar}{c}\vec{B}\cdot\vec{\sigma}\,u - \frac{e\hbar}{ic}(\vec{\sigma}\cdot\vec{\mathcal{E}})v$$

(This part only would yield Klein Gordon equation)

Reduce further neglecting $\frac{1}{c^3}$ terms

(15) $E = mc^2 + w$. Then second (13) gives in lowest approx.

$$(16)\qquad v \approx \frac{1}{2mc}\sigma\cdot p\,u \quad \text{(good enough for this)}$$

(14) becomes:

(use $(\sigma\cdot\mathcal{E})(\sigma\cdot p) = \mathcal{E}\cdot p + i\sigma\cdot\mathcal{E}\times p$)

$$(17)\qquad wu = \mathcal{H}u$$

$$(18)\quad \mathcal{H} = \frac{1}{2m}\left(\vec{p}-\frac{e\vec{A}}{c}\right)^2 + e\varphi - \frac{1}{8m^3c^2}\left(p-\frac{eA}{c}\right)^4 - \frac{e\hbar}{4im^2c^2}\vec{\mathcal{E}}\cdot\vec{p} - \frac{e\hbar}{4m^2c^2}\vec{\sigma}\cdot\vec{\mathcal{E}}\times\vec{p} - \frac{e\hbar}{2mc}\vec{B}\cdot\vec{\sigma}$$

First two terms are classical hamiltonian. Next two terms are spin independent relativ. corrections. The interesting terms are the last two:

(19) $$-\frac{e\hbar}{2mc}\vec{\sigma}\cdot\vec{B}$$

Is energy of mag. mom $\frac{e\hbar}{2mc}\vec{\sigma} = \mu_0\vec{\sigma}$ in mag. field $\underline{B}$.

(20) $$-\frac{e\hbar}{4m^2c^2}\vec{\sigma}\cdot\vec{\mathcal{E}}\times\vec{p}$$

is the mutual energy of $\mu_0\vec{\sigma}$ in apparent magn. field $\vec{\mathcal{E}}\times\frac{\vec{v}}{c} \approx \frac{1}{mc}\vec{\mathcal{E}}\times\vec{p}$ divided by 2 (Thomas correction) See Lect. 26

36 - Dirac Electron in Central field - Hydrogen atom

Assume

(1) $$\varphi = \varphi(r) \qquad \vec{A} = 0$$

(26-(9)) →

(2) $$H = -e\,\varphi(r) + mc^2\beta + c\,\vec{\alpha}\cdot\vec{p}$$

(26-(13)) →

(3) $$\begin{cases} \frac{1}{c}\left(E - mc^2 + e\varphi\right) u = \vec{\sigma}\cdot\vec{p}\, v \\ \frac{1}{c}\left(E + mc^2 + e\varphi\right) v = \vec{\sigma}\cdot\vec{p}\, u \end{cases}$$

Formulas written for electron of charge $-e$

ang. mom (34-(35))

(4) $$\hbar\vec{J} = \vec{x}\times\vec{p} + \frac{\hbar}{2}\vec{\sigma}'$$

Commutes with H, Take then

(5) $$\begin{cases} \vec{J}^2 = j(j+1) \quad \text{and} \\ J_z = m \qquad -j \le m \le j \end{cases}$$

diagonal

Observe $\vec{\sigma}'$ has same commutation properties of $\vec{\sigma}$

(6) $$\sigma_x'^2 = \sigma_y'^2 = \sigma_z'^2 = 1 \qquad \vec{\sigma}'\times\vec{\sigma}' = 2i\vec{\sigma}'$$

Then from (4) + (5) allowable values of $\vec{l}$, l_z are

(7) $$l = j \pm \tfrac{1}{2} \quad + \quad l_z = m \pm \tfrac{1}{2}$$

From (3) follows (because $\vec{\sigma}\cdot\vec{p}$ is a pseudoscalar) that u, v have opposite parity. From this

~~expression~~ find as on p. 26-5 two types of solutions.

First type $(l = j - \frac{1}{2})$

$$(8)\begin{cases} u = \dfrac{R(r)}{\sqrt{2j}} \begin{vmatrix} \sqrt{j+m}\; Y_{j-\frac{1}{2},m-\frac{1}{2}} \\ \sqrt{j-m}\; Y_{j-\frac{1}{2},m+\frac{1}{2}} \end{vmatrix} = R(r)\, Z_{j,j-\frac{1}{2},m} \begin{matrix} \leftarrow 1st \\ \leftarrow 2nd \end{matrix} \\ v = \dfrac{iS(r)}{\sqrt{2(j+1)}} \begin{vmatrix} +\sqrt{j+1-m}\; Y_{j+\frac{1}{2},m-\frac{1}{2}} \\ -\sqrt{j+1+m}\; Y_{j+\frac{1}{2},m+\frac{1}{2}} \end{vmatrix} = iS(r)\, Z_{j,j+\frac{1}{2},m} \begin{matrix} \leftarrow 3rd \\ \leftarrow 4th \end{matrix} \end{cases}$$

Dirac components

Properties of the $Z_{j,j\pm\frac{1}{2},m}$ dicotomic functions

These functions play the role of the spherical harmonics for problems with spin. They have $l = j \pm \frac{1}{2}$

$$(9)\qquad (\vec{\sigma}\cdot\vec{x})\left(f(r)\, Z_{j,j\pm\frac{1}{2},m}\right) = r f(r)\, Z_{j,j\mp\frac{1}{2},m}$$

$$(10)\qquad (\vec{\sigma}\cdot\vec{p})\left(f(r)\, Z_{j,j\pm\frac{1}{2},m}\right) = \frac{\hbar}{i}\left(f'(r) + \left(1 \pm j \pm \tfrac{1}{2}\right)\frac{f}{r}\right) Z_{j,j\mp\frac{1}{2},m}$$

Substituting (8) in (3)

$$(11)\begin{cases} \dfrac{1}{\hbar c}\left(E - mc^2 + e\varphi\right) R(r) = S'(r) + \left(j + \frac{3}{2}\right) S(r)/r \\ \dfrac{1}{\hbar c}\left(E + mc^2 + e\varphi\right) S(r) = -R'(r) + \left(j - \frac{1}{2}\right) R(r)/r \end{cases}$$

The two first order eqns (11) are the equivalent of the single non relativistic radial eq.n of the second order. In this solution

R large
S small $\quad l = j - \frac{1}{2}$

Another type solution has $l = j + \frac{1}{2}$. For it (8) + (11) are instead

<u>Second type</u> $\left(l = j + \frac{1}{2} \right)$

$$(12) \begin{cases} u = R(r)\, Z_{j,j+\frac{1}{2},m} \\ v = -i S\, Z_{j,j-\frac{1}{2},m} \end{cases}$$

And the two coupled radial equations are instead of (11)

$$(13) \begin{cases} \dfrac{E - mc^2 + e\varphi}{\hbar c} R = -S' + \left(j - \frac{1}{2}\right) S/r \\[2ex] \dfrac{E + mc^2 + e\varphi}{\hbar c} S = R' + \left(j + \frac{3}{2}\right) R/r \end{cases}$$

For the Coulomb potential $e\varphi = \frac{Ze^2}{r}$

(11) + (13) can be solved exactly (See Schiff Sect. 44)

For example: ground state of hydrogen-like atom $j=\frac{1}{2}$, $l=0$ (Use first type (8) (11)) (11) are

$$(14)\begin{cases}\left(\varepsilon-\mu+\frac{z}{r}\right)R = S'+\frac{2}{r}S\\ \left(\varepsilon+\mu+\frac{z}{r}\right)S = -R'\end{cases}$$

$$(15)\left\{\varepsilon = \frac{E}{\hbar c}\qquad \mu=\frac{mc}{\hbar}\qquad z=\frac{Ze^2}{\hbar c}=\frac{Z}{137}\right.$$

Try $\qquad R = r^{\gamma}e^{-\lambda r}$

Substituting in (14) find solution with

$$(16)\begin{cases}\gamma=-1+\sqrt{1-z^2}\qquad \lambda = z\mu = Z\,\frac{em}{\hbar^2}\\ \frac{S(r)}{R(r)}=\frac{1-\sqrt{1-z^2}}{z}=\text{constant}\end{cases}$$

$$(17)\begin{cases}\varepsilon=\mu\sqrt{1-z^2}\ \text{ or }\ E=mc^2\sqrt{1-\left(\frac{Ze^2}{\hbar c}\right)^2}=\\ \qquad = mc^2-\frac{Z^2e^4m}{2\hbar^2}-\frac{Z^4e^8}{8\hbar^4c^2}+\dots\end{cases}$$

(mc^2: This is rest energy; $\frac{Z^2e^4m}{2\hbar^2}$: This is non relativistic value)

Normalized solution is

$$(18)\begin{cases}R(r)=(2z\mu)^{\sqrt{1-z^2}}\sqrt{\frac{z\mu\left(1+\sqrt{1-z^2}\right)}{\left(2\sqrt{1-z^2}\right)!}}\ r^{-1+\sqrt{1-z^2}}\ e^{-z\mu r}\\ S(r)=\frac{1-\sqrt{1-z^2}}{z}R(r)\end{cases}$$

Substitute these in (8) with $j=\frac{1}{2}$, $m=\pm\frac{1}{2}$ to find the two normalized ground state solutions with electron spin up or down

37 – Transformation of Dirac spinors.

Rewrite (35-(7)) Dirac eq.ns

(1) $\left(\frac{mc}{\hbar} + \underset{\sim}{\gamma}\cdot\underset{\sim}{\nabla} - \frac{ie}{\hbar c}\underset{\sim}{\gamma}\cdot\underset{\sim}{A}\right)\psi = 0$

Indep. of frame requires: In new frame

(2) $x_\mu \to x'_\mu = a_{\mu\nu} x_\nu$ (Sum over equal indices)

(3) $\psi \to \psi' = T\psi$ (T is 4×4 Dirac-like matrix)

(4) $\begin{cases} \nabla_\mu \to \nabla'_\mu = a_{\mu\nu}\nabla_\nu \\ A_\mu \to A'_\mu = a_{\mu\nu} A_\nu \end{cases}$ ($a_{\mu\nu}$ is orthogonal)

In new frame same eq. for ψ', ∇', A'

$\left(\frac{mc}{\hbar} + \underset{\sim}{\gamma}\cdot\underset{\sim}{\nabla}'\right)\psi' = 0$ (omit A for brevity)

$T^{-1}\psi$ — multiply left by T & find

$\left(\frac{mc}{\hbar} + T\underset{\sim}{\gamma}T^{-1}\cdot\nabla'\right)\psi = 0$

This must be = (1) without A term, which requires

(5) $\boxed{T\gamma_\mu T^{-1} = a_{\mu\nu}\gamma_\nu}$

Consider infinitesimal transformation

(6) $a_{\mu\nu} = \delta_{\mu\nu} + \varepsilon_{\mu\nu}$ neglect squares of ε's

Orthogonality requirement

(7) $\varepsilon_{\mu\nu} = -\varepsilon_{\nu\mu}$ ($\varepsilon_{\nu\nu} = 0$)

(8) $\begin{cases} \text{Reality requirement: } \varepsilon_{nm} \text{ are real} \\ \varepsilon_{4n} = -\varepsilon_{n4} \text{ are pure imag.} \quad n, m = 1,2,3 \end{cases}$

Assume T differs from unit matrix by order ε

(9) $$T = 1 + S$$ (S order ε)

then

(10) $$T^{-1} = 1 - S$$

and (5) →

(11) $$S\gamma_\mu - \gamma_\mu S = \varepsilon_{\mu\nu}\gamma_\nu$$

This condition is satisfied by

(12) $$S = -\frac{1}{4}\varepsilon_{\mu\nu}\gamma_\mu\gamma_\nu$$

Therefore

(13) $$T = 1 - \frac{1}{4}\sum_{\mu\nu}\varepsilon_{\mu\nu}\gamma_\mu\gamma_\nu$$

Lorentz group combined from infinitesimal transformations (6) on coordinates (13) on ψ

Example: infinitesimal rotation around z

(14) $$x'_4 = x_4 \quad x'_3 = x_3 \quad x'_1 = x_1 - \varepsilon x_2 \quad x'_2 = x_2 + \varepsilon x_1$$

or $\varepsilon_{12} = -\varepsilon$ $\varepsilon_{21} = \varepsilon$ all others zero

$$T_\varepsilon = 1 + \frac{\varepsilon}{2}\gamma_1\gamma_2 = \begin{vmatrix} 1+\frac{i}{2}\varepsilon & 0 & 0 & 0 \\ 0 & 1-\frac{i}{2}\varepsilon & 0 & 0 \\ 0 & 0 & 1+\frac{i\varepsilon}{2} & 0 \\ 0 & 0 & 0 & 1-\frac{i\varepsilon}{2} \end{vmatrix}$$

For finite rotation around z by angle φ (take $T_\varepsilon^{\varphi/\varepsilon} = T_\varphi$) find:

(15) $$T_\varphi = \begin{vmatrix} e^{\frac{i\varphi}{2}} & 0 & 0 & 0 \\ 0 & e^{-\frac{i\varphi}{2}} & 0 & 0 \\ 0 & 0 & e^{\frac{i\varphi}{2}} & 0 \\ 0 & 0 & 0 & e^{-\frac{i\varphi}{2}} \end{vmatrix}$$

Corresp. transformation of ψ

(16) $\psi_1' = e^{i\frac{\varphi}{2}}\psi_1 \quad \psi_2' = e^{-\frac{i\varphi}{2}}\psi_2 \quad \psi_3' = e^{\frac{i\varphi}{2}}\psi_3 \quad \psi_4' = e^{-\frac{i\varphi}{2}}\psi_4$

Observe: for $\varphi = 2\pi \quad \psi' = -\psi$ (Comments)

Example: Infinitesimal Lorentz transform

$$(17) \quad \begin{array}{ll} x_1' = x_1 - \varepsilon t c = x_1 + i\varepsilon x_4 & x_2' = x_2 \\ x_4' = x_4 - i\varepsilon x_1 & x_3' = x_3 \end{array}$$

$$(18) \quad T_\varepsilon = 1 - \frac{i\varepsilon}{2}\gamma_1\gamma_4 = 1 + \frac{\varepsilon}{2}\alpha_1 = \begin{Vmatrix} 1 & 0 & 0 & \frac{\varepsilon}{2} \\ 0 & 1 & \frac{\varepsilon}{2} & 0 \\ 0 & \frac{\varepsilon}{2} & 1 & 0 \\ \frac{\varepsilon}{2} & 0 & 0 & 1 \end{Vmatrix}$$

Obtain finite Lorentz transf.

($x_0 = ct$)

$$(19) \quad x_1' = \frac{x_1 - \beta x_0}{\sqrt{1-\beta^2}} \qquad x_0' = \frac{x_0 - \beta x_1}{\sqrt{1-\beta^2}}$$

by iterating (17) a number of times

$$n = \frac{1}{\varepsilon}\,\mathrm{artgh}\,\beta$$

Take corresp

(because $\alpha_1^2 = 1$)

$$(20) \quad \begin{cases} T_\beta = T_\varepsilon^{\,n} = \left(1 + \frac{\varepsilon}{2}\alpha_1\right)^n = e^{\frac{n\varepsilon}{2}\alpha_1} = \\ = \cosh\frac{n\varepsilon}{2} + \alpha_1 \sinh\frac{n\varepsilon}{2} = \\ = \cosh\left(\frac{1}{2}\mathrm{artgh}\,\beta\right) + \alpha_1 \sinh\left(\frac{1}{2}\mathrm{artgh}\,\beta\right) = \\ = \sqrt{\frac{1+\sqrt{1-\beta^2}}{2\sqrt{1-\beta^2}}} + \alpha_1\sqrt{\frac{1-\sqrt{1-\beta^2}}{2\sqrt{1-\beta^2}}} \end{cases}$$

Space reflection

$$(21)\quad \begin{cases} x'_n = -x_n & n = 1,2,3 \\ x'_4 = x_4 \end{cases}$$

$$(22)\quad \{\psi \to \psi' = T_{ref}\,\psi$$

From (5)

$$(23)\quad T_{ref}\,\gamma_n\,T_{ref}^{-1} = -\gamma_n\,,\quad T_{ref}\,\gamma_4\,T_{ref}^{-1} = \gamma_4$$

Satisfied by

$$(24)\quad \boxed{T_{ref} = \gamma_4 = \beta}$$

Observe:

$$(25)\quad T_{ref} = T_{ref}^{-1} = \tilde{T}_{ref}$$

Observe: for our choice of γ_4 (34-(10))

$$(26)\quad \psi'_1 = \psi_1 \quad \psi'_2 = \psi_2 \quad \psi'_3 = -\psi_3 \quad \psi'_4 = -\psi_4$$

Parity behavior change between ψ_1, ψ_2 and ψ_3, ψ_4. Then: for an even state

$$(27)\quad \begin{cases} \psi_1(\vec{x}) = \psi_1(-\vec{x})\,,\ \psi_2(\vec{x}) = \psi_2(-\vec{x}),\ \psi_3(\vec{x}) = -\psi_3(-\vec{x}),\ \psi_4(\vec{x}) = -\psi_4(-\vec{x}) \\ \text{and for an odd state} \\ \psi_{1,2}(\vec{x}) = -\psi_{1,2}(\vec{x})\ ;\ \psi_{3,4}(\vec{x}) = \psi_{3,4}(-\vec{x}) \end{cases}$$

Compare with (36-(8)(12)). Find: parity of l = parity of state for electron states. For positron states the large components are ψ_3, ψ_4 which have parity reversed.

Properties

$$(28)\quad T_{ref}\,\gamma_\mu\,\tilde{T}_{ref} = \begin{cases} -\gamma_\mu \text{ for } \mu = 1,2,3 \\ \gamma_\mu \text{ for } \mu = 4 \end{cases} \text{ and } T_{ref}\,\beta\gamma_\mu\,\tilde{T}_{ref} = \begin{cases} -\beta\gamma_\mu & (\mu = 1,2,3) \\ \beta\gamma_\mu & (\mu = 4) \end{cases}$$

Dirac spinor operators as scalars, vectors, tensors.

From (8) (13) (latin indices = 1,2,3; greek indices = 1,2,3,4; sum over equal indices)

(29)
$$T = 1 - \frac{1}{4}\varepsilon_{\mu\nu}\gamma_\mu\gamma_\nu = 1 - \frac{1}{4}\varepsilon_{mn}\gamma_m\gamma_n - \frac{1}{2}\varepsilon_{4n}\beta\gamma_n$$

($\beta = \gamma_4$) ($\gamma_\mu\gamma_\nu + \gamma_\nu\gamma_\mu = 0$) ($\varepsilon_{mn}$ real) (ε_{4n} imaginary)

$$T^{-1} = 1 + \frac{1}{4}\varepsilon_{\mu\nu}\gamma_\mu\gamma_\nu = 1 + \frac{1}{4}\varepsilon_{mn}\gamma_m\gamma_n + \frac{1}{2}\varepsilon_{4n}\beta\gamma_n$$

$$\tilde{T} = 1 + \frac{1}{4}\varepsilon^*_{\mu\nu}\gamma_\mu\gamma_\nu = 1 + \frac{1}{4}\varepsilon_{mn}\gamma_m\gamma_n - \frac{1}{2}\varepsilon_{4n}\beta\gamma_n$$

(30) In general $\tilde{T} \neq T^{-1}$ (T non unitary: (comments))
T is unitary when $\varepsilon_{4n} = 0$ (i.e. pure space rotation)

Finds

(31)
$$\beta\tilde{T}\beta = T^{-1}$$
$$\tilde{T}\beta = \beta T^{-1}, \quad \beta\tilde{T} = T^{-1}\beta$$

ⓐ Search for spinor matrices u behaving as a scalar. Means: for frame change (2)

$$x_\mu \to x'_\mu = a_{\mu\nu}x_n$$ and associated spinor change
$$\psi \to \psi' = T\psi$$

(32) The expression $\tilde{\psi} u \psi \to \tilde{\psi}' u \psi' = \tilde{\psi} u \psi$

$$\tilde{\psi}' u \psi' = \widetilde{T\psi}\, u T\psi = \tilde{\psi}\,\tilde{T} u T\psi$$

Then should be

$$\tilde{T} u T = u$$

$$\tilde{T} u T = \beta T^{-1}\beta u T = u \quad \text{hence} \quad (\beta^2 = 1)$$

$$(\beta u) T = T(\beta u) \quad \text{satisfied for } T = (29)$$

(33) by $\beta u = 1$ and $\beta u = \gamma_1\gamma_2\gamma_3\gamma_4 = \gamma_5$

Two soln's

$$u = \beta 1 \quad \text{and} \quad u = \beta\gamma_5$$

behave differently for space reflection $T_{ref} = \beta$

$$\tilde{T}_{ref}\, \beta 1\, T_{ref} = \beta\beta 1 \beta = 1\beta = \beta 1$$

$$\tilde{T}_{ref}\, \beta\gamma_5\, T_{ref} = \beta\beta\gamma_5\beta = \gamma_5\beta = -\beta\gamma_5$$

Therefore:

$\beta 1$ = scalar ← $\tilde{\psi}\beta 1 \psi$

$\beta\gamma_5$ = pseudoscalar ← $\tilde{\psi}\beta\gamma_5\psi$

Comments on β- factor (notation)

$\psi^{+} = \tilde{\psi}\beta$ Then

(34)
$\psi^{+} 1 \psi$ transforms like a scalar
$\psi^{+}\gamma_5\psi$ " " a pseudoscalar

Comment: pseudoscalar pion interaction term $\varphi\psi^{+}\gamma_5\psi$ of field theory

Other Dirac operators are such that $\psi^{+}u_\mu\psi$ or $\psi^{+}u_{\mu\nu}\psi$ transform like the components of ~~pseudovector~~ four vectors axial four vectors or antisymmetric tensor.

This means e.g. $\psi'^{+}u'_\mu\psi' = \psi^{+}u_\mu\psi$ with $u'_\mu = a_{\mu\nu}u_\nu$ (see (24))

(35)

1	scalar
γ_5	pseudoscalar
$\gamma_1, \gamma_2, \gamma_3, \gamma_4$	four vector
$\gamma_2\gamma_3\gamma_4,\ \gamma_3\gamma_1\gamma_4,\ \gamma_1\gamma_2\gamma_4,\ \gamma_1\gamma_2\gamma_3$	axial four vector
$\gamma_2\gamma_3,\ \gamma_3\gamma_1,\ \gamma_1\gamma_2,\ \gamma_1\gamma_4,\ \gamma_2\gamma_4,\ \gamma_3\gamma_4$	antisym. tensor

Observe: all spinor operators are linear combinations of the 16 below

Time reversal - (General comments)

$$(36)\quad \begin{cases} \vec{x} \to \vec{x} & \vec{A} \to -\vec{A} & \vec{\nabla} \to \vec{\nabla} \\ \vec{x}_4 \to -\vec{x}_4 & A_4 \to A_4 & \nabla_4 \to -\nabla_4 \end{cases}$$

Then $\exists$ ψ a solution of (1)

$$(37)\quad 0 = \frac{mc}{\hbar}\psi + \vec{\gamma}\cdot\left(\vec{\nabla} - \frac{ie}{\hbar c}\vec{A}\right)\psi + \gamma_4\left(\frac{\partial}{\partial x_4} - \frac{ie}{\hbar c}A_4\right)\psi$$

The corresp. time reversed solution ψ' must solve time reversed eq'n of (37)

$$(38)\quad 0 = \frac{mc}{\hbar}\psi' + \vec{\gamma}\cdot\left(\vec{\nabla} + \frac{ie}{\hbar c}\vec{A}\right)\psi' - \gamma_4\left(\frac{\partial}{\partial x_4} + \frac{ie}{\hbar c}A_4\right)\psi'$$

Clearly impossible to solve with $T\psi$.

However

$$(39)\quad \psi' = S\psi^*$$

may work. From (37) $(i \to -i)$

$$(40)\quad 0 = \frac{mc}{\hbar}\psi^* + \vec{\gamma}^*\cdot\left(\vec{\nabla} + \frac{ie}{c\hbar}\vec{A}\right)\psi^* - \gamma_4^*\left(\frac{\partial}{\partial x_4} + \frac{ie}{\hbar c}A_4\right)\psi^*$$

Multiply to left by S. Identify to (38). Required

$$(41)\quad S\vec{\gamma}^*S^{-1} = \vec{\gamma} \qquad S\gamma_4^*S^{-1} = -\gamma_4 \qquad \psi' = S\psi^*$$

(41) can be fulfilled e.g. for Standard form (34-(9)(10)) of γ's by

$$(42)\quad S = i\gamma_1\gamma_3 = \begin{vmatrix} 0 & -i & 0 & 0 \\ i & 0 & 0 & 0 \\ 0 & 0 & 0 & -i \\ 0 & 0 & i & 0 \end{vmatrix} = \sigma'_y$$

see (34-(36))

Charge conjugation. General comments.
Solutions of (37) contain both electron + positron sol'ns. Then expect that from each solution ψ it should be possible to obtain a ψ' obeying (37) with

$$e \to -e \tag{43}$$

$$\frac{mc}{\hbar}\psi' + \vec{\gamma}\cdot\left(\vec{\nabla} + \frac{ie}{\hbar c}\vec{A}\right)\psi' + \gamma_4\left(\nabla_4 + \frac{ie}{\hbar c}A_4\right)\psi' = 0 \tag{44}$$

Try transform

$$\psi' = C\psi^* \tag{45}$$

Apply C to left of compl. conj eq.n (40). Find that it goes into (44) provided:

$$C\vec{\gamma}^* C^{-1} = \vec{\gamma}\ , \quad C\gamma_4^* C^{-1} = -\gamma_4 \tag{46}$$

For standard form of γ's (34-p 3)
Solution of (46) is $C = \gamma_2$
Charge conjugate solution is

$$\boxed{\psi_{ch.conj} = \gamma_2\,\psi^*} \tag{47}$$

PROBLEMS

for

NOTES ON QUANTUM MECHANICS

INTRODUCTION

At the end of each lecture Fermi would always make up a problem, which was usually closely related to what he had just talked about that day. He did not appear to have prepared the problems in advance. He conveyed the strong impression that he regarded working these problems an essential part of the course. It seemed that, in Fermi's view, the power of the intellectual edifice of physics is manifested in quantitative results, correct as calculations and verifiable in experiment. Accordingly, most problems call for numerical answers; formal proofs are rare.

Often Fermi specified a graphical presentation by which he meant a complete, easily understood drawing, to scale if possible. Occasionally, as in problems 12 and 58, he called for an experimental design, which was to be worked out quantitatively with regard to relevant magnitudes, and which required the student to find published papers and other reference information. He rarely gave references himself. For further detail on some subjects, Fermi occasionally mentioned Leonard Schiff's book, *Quantum Mechanics*, First Edition.

Fermi did not use lecture time to go through problems. However, he came to his office reliably at 8:00 A.M. and was available to students. When I took these courses—two quarters of approximately sixty lectures and sixty problems—the problems were read by Lee C. Teng, a Ph.D. student of Gregor Wentzel. Teng relates that Fermi wanted a complete copy of the given problems and solutions for his file, as well as for the record of each student.

Robert A. Schluter

PROBLEMS

Each problem is identified with the section title and page number of the notes to which it is related. Although equations and derivations from the notes are duplicated as necessary for the reader's convenience, the original equation numbers are provided so that the reader may easily cross-reference the equations within the context of the notes.

Schrödinger equation, p. 4

1. For $f(t)$ given, find the development in monochromatic waves of form

$$\psi = ue^{-i\omega t} = ue^{-(1/\hbar)wt} \text{ (Eq. 2)}$$

a) When the wave repeats:

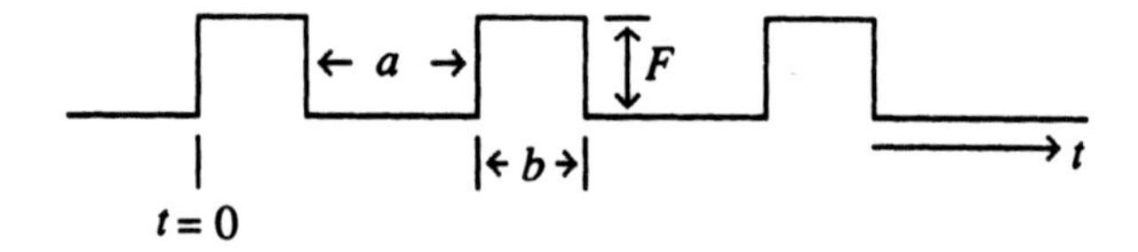

b) When the wave consists of one pulse:

Simple one-dimensional problems, p. 7

2. Confirm by direct calculation the normalization factors given in the following:

a) For a free particle on closed line of length a, where $U(x) = 0$ and

$$u \sim e^{\pm i\left(2mE/\hbar^2\right)^{1/2}x} \text{ (Eq. 2),}$$

the periodicity condition requires that $u \sim e^{(2\pi i/a)lx}$, where l is an integer. Therefore,

$$E_l = \frac{2\pi^2\hbar^2}{ma^2}l^2 \text{ (Eq. 3)}$$

and the normalized functions are

$$u_l = a^{-1/2}e^{(2\pi il/a)x} \text{ (Eq. 4).}$$

b) For a rotator with fixed axis, as above with $m \to A$ = moment of inertia, $a \to 2\pi$, $x \to \alpha$,

$$\begin{cases} E_l = \left(\hbar^2 / 2A\right) l^2 \\ u_l = (2\pi)^{-1/2} e^{il\alpha} \end{cases} \quad \text{(Eq. 5)}$$

3. Solve the one-dimensional Schrödinger equation

$$u'' + \frac{2m}{\hbar^2}(E - U)u = 0 \text{ (Eq. 1)}$$

for the potential $U(x)$, given below. The particle is an electron where $A = 10$ eV, $a = 10^{-8}$ cm. Consider the cases $E > A$ and $E < A$.

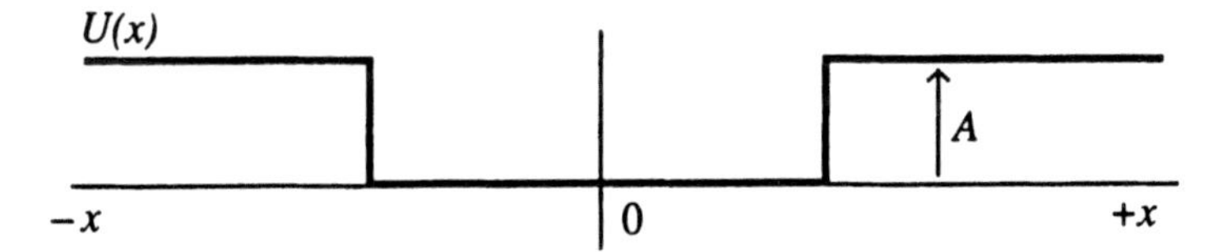

Simple one-dimensional problems, p. 8

4. For a point on an infinite line (zero potential)

$$u'' + (2mE / \hbar^2)u = 0 \text{ (Eq. 8)}$$

consider the "alternate approach" using a wave packet form

$$u_{\delta k} = \int_{k_0-(\delta k/2)}^{k_0+(\delta k/2)} e^{ikx}\, dk .$$

Normalize this.

Linear oscillator, p. 11

5. Make careful graphs of the oscillator eigenfunctions

$$u_n = (m\omega / \hbar)^{1/4} \left(\sqrt{\pi}\, 2^n n!\right)^{-1/2} H_n(\xi) e^{-\xi^2/2}, \text{ where } \xi = (m\omega / \hbar)^{1/2} x \text{ (Eq. 17)},$$

for $n = 0, 1, 2,$ and 3.

6. Given a function $f(\xi) = e^{-2\xi^2}$,

a) Find the coefficients in the expansion of this function in terms of the oscillator eigenfunctions in problem 5, Eq. 17, and

b) Add graphically the first several terms in this expansion and compare (graphically) with $f(\xi)$.

W. K. B. method, pp. 12–14

7. Consider the one-dimensional oscillator wave function for $n = 7$. Compare the W. K. B. solution with the exact solution. Plot both together.

Spherical harmonics, p. 16 (also, Hydrogen atom, p. 21)

8. Consider the function $e^{-(r^2/a^2)}$ where a is the Bohr radius and r is distance from the origin. Develop this in eigenfunctions of the hydrogen problem. Determine coefficients for 1s, 2s, 3s, and all $p, d, f, \ldots$ states.

Central forces, p. 17

9. Find the lowest s-state energy level for an electron in the field of a spherically symmetrical potential $U = -10$ eV for $r < 10^{-8}$ cm and $U = 0$ for $r > 10^{-8}$ cm (answer: $E \sim -.032$ eV).

10. An electron is in a central field potential $U = f(r)$. Find the lowest s and p states.

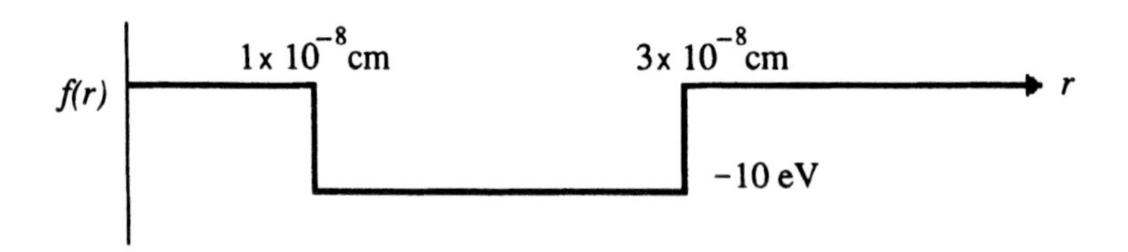

Central forces, p. 18

11. Treat the HCl molecule like a rigid dumbbell and find the energy levels (in eV units) and plot. The HCl molecule consists of masses 35.0 and 1.0 at an H to Cl distance 1.27×10^{-8} cm. Compare to experimental data.

Hydrogen atom, p. 20

12. Spectroscopes are available having optical parts made of fluorite, quartz, glass, and rock salt. Using

$$E_n = -\frac{mZ^2e^4}{2\hbar^2u^2}, \quad \text{where } n = l+1, l+2, \ldots \text{ (Eq. 11)},$$

determine which optical component(s) should be used to observe the lines of hydrogen.

Hydrogen atom, pp. 21–22

13. Using Eq. 20:

$$u(1s) = \frac{1}{\sqrt{\pi a^3}} e^{-r/a}$$

$$u(2s) = \frac{[2-(r/a)]e^{-r/2a}}{4\sqrt{2\pi a^3}}$$

$$u(2p) = \frac{(r/a)e^{-r/2a}}{8\sqrt{\pi a^3}} \begin{cases} -\sin\vartheta e^{i\varphi} \\ \sqrt{2}\cos\vartheta \\ \sin\vartheta e^{-i\varphi} \end{cases}$$

map out (graph) the eigenfunctions for hydrogen-like states $1s$, $2s$, $2p_0$, and $3s$.

Orthogonality of wave functions, p. 27

14. Given $\psi_1 = e^{-x^2}, \psi_2 = e^{-2x^2}, \psi_3 = e^{-3x^2}$, all for one eigenvalue, find an orthonormal basis, ϕ_1, ϕ_2, ϕ_3.

Linear operators, p. 28

15. D is the operator d/dx. Evaluate

$$[D, x^2], \; [D, x^3], \; [D, x^4], \; [D, x^n], \; [D^2, x^2].$$

Linear operators, pp. 28–31

16. These operators are important in spin theory:

$$A = \begin{pmatrix} 0 & 1 \\ 1 & 0 \end{pmatrix}, \quad B = \begin{pmatrix} 0 & -i \\ i & 0 \end{pmatrix}, \quad C = \begin{pmatrix} 1 & 0 \\ 0 & -1 \end{pmatrix}.$$

Confirm, in two ways, that

$$A^2 = 1, \qquad B^2 = 1, \qquad C^2 = 1,$$

$$AB = -AB, \quad BC = -CB, \quad CA = -AC.$$

Find $[A,B]$, $[B,C]$, and $[C,A]$ in terms of A, B, and C.

17. Find the eigenvalues of

$$\begin{pmatrix} 0 & 1 \\ 1 & 0 \end{pmatrix}, \begin{pmatrix} 0 & -i \\ i & 0 \end{pmatrix}, \begin{pmatrix} 1 & 0 \\ 0 & -1 \end{pmatrix}, \begin{pmatrix} 0 & \sqrt{2} & 0 \\ \sqrt{2} & 0 & \sqrt{2} \\ 0 & \sqrt{2} & 0 \end{pmatrix},$$

$$\begin{pmatrix} 2 & 0 & 0 \\ 0 & 3 & 0 \\ 0 & 0 & -1 \end{pmatrix}, \begin{pmatrix} b & 0 & 0 & 0 \\ 0 & q & 0 & 0 \\ 0 & 0 & r & 0 \\ 0 & 0 & 0 & s \end{pmatrix}.$$

Linear operators, pp. 28–31 (also, Eigenvalues and eigenfunctions, pp. 32–36)

18. Show by calculation that $G\,f(x) = 0\,f(x)$ for any $f(x) = \rho(x)e^{i\theta(x)}$ if

$$G = \left(p - \hbar\frac{d\theta}{dx}\right)^2 + \frac{\hbar^2}{\rho}\frac{d^2\rho}{dx^2}.$$

Operators for mass point, Postulates, pp. 39–40

19. The energy operator for the linear oscillator is

$$A = \frac{p^2}{2m} + \frac{m\omega^2}{2}x^2.$$

Eigenvalues are

$$\hbar\omega(n + 1/2).$$

Suppose a measurement yields $\hbar\omega/2$. Immediately after this measurement, another measurement is made. Find the probability that it yields $3\hbar\omega'/2$, where this is an eigenvalue of

$$\frac{p^2}{2m}+\frac{m\omega'^2}{2}x^2,$$

the operator for an oscillator of classical frequency ω'.

20. Develop e^{-x^2+i7x} in a series of eigenfunctions of p, the momentum. Plot this function, and plot the amplitude of the "coefficient" of the eigenfunctions of p.

Uncertainty principle, p. 45

21. Take the wave function $\psi_{(x)} = e^{-k^2x^2}$. Determine $\bar{x}$, $\bar{p}$, Δx, and Δp. Verify

$$\Delta x\, \Delta p \geq \hbar/2.$$

(Fermi defined $\Delta x \equiv \left(\overline{(x-\bar{x})^2}\right)^{1/2}$, where $\bar{x}$ denotes the quantum mechanical mean value, p. 40.)

Hermitian matrices—eigenvalue problems, p. 55 (also, Linear operators, p. 31, and Conservation theorems, p. 87)

22. R is the operation defined by $R\, f(x) = f(-x)$. Is R hermitian? What are its eigenvalues and eigenfunctions? Answer the same questions for the displacement operator

$$e^{aD} f(x) = f(x+a).$$

Hermitian matrices—eigenvalue problems, p. 58

23. Find the eigenvalues and eigenfunctions of and discuss the ellipsoid associated with

$$\begin{pmatrix} 0 & 1 & 0 \\ 1 & 5 & 1 \\ 0 & 1 & 0 \end{pmatrix}.$$

The angular momentum, pp. 74–76

24. The hamiltonian for a particle in a central field of arbitrary dependence on r is

$$H = \frac{p^2}{2m} + U(r).$$

Prove that H and M^2 commute and that H and M_z commute.

Time-independent perturbation theory—Ritz method, p. 92

25. Referring to

$$H = \overbrace{\frac{1}{2M}p^2 + U(r)}^{H_0} - \overbrace{\frac{eB}{2Mc}\left(xp_y - yp_z\right)}^{\mathcal{H}} \text{ (Eq. 27)},$$

the perturbation term in the hamiltonian for hydrogen in a magnetic field $\boldsymbol{B}$ in the $+z$ direction is diagonal in the usual representation in which angular momentum is quantized along the $+z$ direction. Now assume $\boldsymbol{B}$ is in the $+x$ direction. Solve for the first-order energies. Work out the eigenfunctions and show that they are linear combinations of those resulting with $\boldsymbol{B}$ along $+z$.

26. *Time-independent perturbation theory—Ritz method, p. 94*
Find the minimum energy eigenvalue of the hydrogen problem (Ritz method) for two "guess" functions.

a) Assume $e^{-\alpha r}$. Find α and the exact ground state energy E_0.
b) Assume a Gaussian $e^{-\alpha r^2}$. Find α and E_0.

Time-independent perturbation theory—Ritz method, pp. 89–95

27. A linear oscillator $(p^2/2m) + (m\omega^2/2)x^2$ is perturbed by $V_p = -Fx$.

a) Calculate the first-order perturbation of E for $n = 0$, 1, and 2.
b) For $n = 0$, calculate the first-order perturbation of the eigenfunction.

28. Consider a linear oscillator $H_0 = (p^2/2m) + (m\omega^2/2)x^2$ with unperturbed energies $E_n = \hbar\omega(n + \frac{1}{2})$. It is perturbed by a term $\mathcal{H} = ax^4$.

a) Calculate the first-order perturbation of E_n for $n = 0$, 1, and 2.
b) For $n = 0$, calculate the first-order perturbation of the eigenfunction.

29. Fermi used time-independent perturbation theory to set up the Van der Waals potential between two atoms interacting via the electric dipole-dipole interaction. The Van der Waals term is

$$\frac{K}{r^6}, \quad \text{where } K = \sum_{nm} \frac{\left|2x_{1n}\xi_{1m} - y_{1n}\eta_{1m} - z_{1n}\zeta_{1m}\right|^2}{E_n + E_m - E_1 - F_1}$$

in which $(xyz)_{1n}$ and $(\xi\eta\zeta)_{1m}$ are the matrix elements of the components of the dipole moment for the two systems. Calculate the most important terms in the series for K for two hydrogen atoms. Neglect spin, and calculate for the three $2p$ states.

Case of degeneracy or quasi degeneracy—hydrogen Stark effect, p. 97

30. Fermi used the Stark effect in hydrogen defined on p. 97 as an example of a degenerate first-order perturbation calculation and assigned as homework finishing the problem and obtaining first-order energies and zeroth-order eigenfunctions (see Eq. 14, p. 98).

31. Calculate the perturbation by an electric field F of these hydrogen levels (Stark effect): $1s$, $2s$, $3s$, $2p$, $3p$, and $3d$. Calculate ν_{23}, the H_α line. Look up the experiment in the literature and compare with your results.

Time-dependent perturbation theory—Born approximation, pp. 99–102

32. A system has two states: $\begin{matrix} \text{———} & 1 \\ \text{———} & 2 \end{matrix}$

The perturbation matrix elements $\mathcal{H}_{nm}$ do not contain time explicitly; that is, $\mathcal{H}_{11}$, $\mathcal{H}_{22}$, $\mathcal{H}_{21}$, and $\mathcal{H}_{12}$ are constants. Refer to

$$\dot{a}_s = -\frac{i}{\hbar}\sum_n a_n \langle s|\mathcal{H}|n\rangle e^{(1/\hbar)\left(E_0^{(s)} - E_0^{(n)}\right)t} \quad \text{(Eq. 7, p. 99).}$$

Integrate the equations exactly, without assumptions. Plot the probabilities in time for the two states.

Emission and absorption of radiation, p. 106

33. From the transition rates

$$(n,l,m) \rightarrow (n',l+1,\text{any } m') = \frac{4}{3}\frac{e^2\omega^3}{\hbar c^3}\frac{l+1}{2l+1}\mathfrak{I}^2 \quad \text{(Eq. 21)}$$

and similarly

$$(n,l,m \rightarrow n',l-1,\text{any } m) = \frac{4}{3}\frac{e^2\omega^3}{\hbar c^3}\frac{l}{2l-1}\left\{\int_0^\infty R_{nl}(r)R_{n',l-1}(r)r^3\,dr\right\}^2 \quad \text{(Eq. 22)}$$

for allowed transitions in hydrogen-like atoms, prove the following. Let P^M be the sum of probabilities for all possible transitions from a given value of m and for a particular Δn. Then all P^M are equal.

Emission and absorption of radiation, p. 107.

34. Determine the lifetime of the $3p$ state of hydrogen.

Emission and absorption of radiation, pp. 103–107

35. A hydrogen atom is exposed to the sun's radiation. Assume the sun emits as a blackbody with $T = 6000°$, is 1.5×10^8 km away, and has a diameter of 1.4×10^5 km. The atom is in the $1s$ state. Compute the rate of absorption of energy for transitions to the $2p_0$ state.

Pauli theory of spin, p. 108

36. Find a change of basis (a linear substitution of the spin ψ's) so that

$$\sigma_x \rightarrow \sigma_y, \quad \sigma_y \rightarrow \sigma_z, \quad \sigma_z \rightarrow \sigma_x\,.$$

37. Find the eigenvalues and eigenfunctions of the following operator on Pauli spin:

$$3\sigma_x - \sigma_y + \sigma_z \quad \left(= \sigma\cdot\underline{Q}, \text{ where } \underline{Q} = (3,-1,1)\right).$$

Pauli theory of spin, p. 110

38. A neutron has an intrinsic angular momentum $\hbar/2$, and its magnetic moment is -1.91 nucleon magneton (which is the Bohr magneton/1836.8). The magnetic moment is

$$\mu_{\text{neut}}(\sigma_x,\sigma_y,\sigma_z).$$

A neutron of velocity 2.2×10^5 cm/s in the $+x$ direction crosses a magnetic field of B = 10,000 gauss in the $+y$ direction and has an extent of 10 cm along the x axis. At the beginning the neutron spin is along $+z$ and the spin function is $\binom{1}{0}$.

What is the probability that the spin is again in the $+z$ direction when it leaves the field? (Consider the precision needed in μ_{neut}.)

Pauli theory of spin, pp. 108–110

39. Consider an operator on a field of three points (1, 2, 3): $\begin{pmatrix} 0 & -i & 0 \\ i & 0 & -i \\ 0 & i & 0 \end{pmatrix}$.

a) Determine its eigenvalues and eigenfunctions.

b) Given a state function $\begin{pmatrix} f_1 \\ f_2 \\ f_3 \end{pmatrix} = \begin{pmatrix} 2 \\ 5 \\ i \end{pmatrix}$, find the *mean* value of the operator.

Electron in central field, p. 116

40. Calculate the doublet splitting for the 2*p* levels of hydrogen (answer: 0.368 cm^{-1}).

Electron in central field, pp. 111–117

41. Calculate the *K* ($2p \to 1s$) lines of Fe. Calculate splitting (spin-orbit and relativity) using a one-electron theory.

Electron in central field, p. 116–117

42. Correct completely the H_α line where $n = 3 \to n = 2$. Include spin correction and relativity correction. Work out intercombinations, find the structure of lines and values of $\overline{\nu}$, and find the wave numbers of transitions.

Anomalous Zeeman effect, pp. 118–119

43. Consider the Zeeman effect for the sodium D doublet, wavelengths 5896 and 5890 Å. Assume the field is 10,000 gauss. Draw to scale the splitting of the lines. Note that the selection rule reduces the number of lines from eight to six.

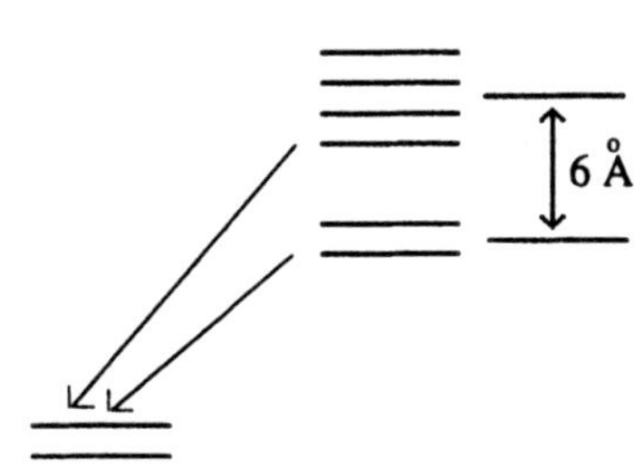

44. Calculate the Zeeman splitting of a triplet level with $L = 2$ and $S = 1$. Before the field is applied, the separation of $J = 3$ and $J = 2$ is $100\ \mathrm{cm}^{-1}$. Draw energy levels for $B = 10{,}000$ gauss.

45. Work out the pattern for the Zeeman effect for a *p-d* combination in a one-electron atom. Assume an arbitrary separation of $\bar{\nu}_0$ between levels, and a weak field. Selection rules: $l \to l+1, l-1;\ j \to j+1, j, j-1$.

46. Consider the p level of a one-electron atom. The doublet separation is $5\ \mathrm{cm}^{-1}$. Plot energy values as a function of the magnetic field strength B. Follow the levels as B increases.

47. Calculate the Zeeman effect for transitions between $L = 0$, $S = 2$, $J = 2$, and $L = 1$, $S = 2$, $J = 1$.

Atomic multiplets, p. 129

48. Draw the term-level diagram for calcium. Look up term values, and plot and label terms. Use selection rules to mark the permitted transitions.

Atomic multiplets, "Spectra of alkalis," p. 129

49. Consider an atom with $Z \geq 10$ in which the effective charge distribution is spherically symmetrical, and make the assumption that the effective potential can be represented by

$$U(r) = -\frac{Ze^2}{r}\left(1 + \frac{\beta}{r}\right),$$

which defines β. Consider s states. The Schrödinger equation gives the relation

$$-\alpha(1-\alpha) = -\frac{2mZe^2\beta}{\hbar^2}$$

and energy levels

$$-E = \frac{mZ^2e^4}{2\hbar^2(n-\alpha)^2}$$

in which the principal quantum number n, an integer, forms with α, a generalized parameter. Look up the energy levels of Na, represent them in a formula (begin with $n = 3$), and determine α, then β, for s states.

50. Plot and label the term values of rubidium (Rb). Mark emission transitions and absorption transitions. Find the α for the Rydberg correction to n in the formula

$$-E_n = \frac{109737}{(n-\alpha)^2}\ \text{cm}^{-1}.$$

Evaluate for the s, p, and d series.

51. Discuss the absorption spectrum of thallium vapor from low temperature to a temperature such that the vapor pressure is ~1 cm Hg.

Atomic multiplets, Atomic electron shells, pp. 129, 130

52. Write out the electron configuration scheme for all 92 elements (p. 130 shows 100 elements, including #100, fermium).

Atomic multiplets, pp. 126–131

53. You have the spectrum of an alkali earth (column II of the periodic table). In the absorption spectrum, you find a low-lying line not part of a multiplet. The problem is to identify this line. To do this, you observe the Zeeman effect. The line splits into three, with separation equal to 3/2 times the *normal* Zeeman line. Place this transition in the level scheme of the atom. (This involves the Landé g factor. Answer: ${}^1S_0 \rightarrow {}^3P_1$.)

Two-electron system, p. 139

54. Draw to scale the term-level diagram for helium, to $n = 5$, showing ortho- and para-diagrams. Include term values.

Hydrogen molecule, Band spectra of diatomic molecules, p. 144

55. Consider the infrared band spectrum of HCl, with fixed separation r_{AB}. Assume that there are no electronic transitions. Fit experimental data to an equation for the energy of the form

$$E_{\text{TOTAL}} = E_{\text{elect}} + \hbar\omega\left(v + \frac{1}{2}\right) + \frac{\hbar^2}{2I}\left(J(J+1) - \Omega^2\right)$$

in the approximation that I, the moment of intertia, ω, the vibrational parameter, and v, the vibrational quantum number are all constant. Find by fitting to experimental data values for I, ω, Ω, and r_{AB}.

56. Note that the frequency emitted or absorbed in rotational transitions (for a particular electronic and vibrational transition) can be given in the form:

$$\bar{\nu} = A + \bar{\nu}' v' - \bar{\nu} v + B' J'(J'+1) - BJ(J+1),$$

where $J' = J + 1$ gives the R branch, $J' = J$ the P branch, and $J' = J - 1$ the Q branch. Consider the diatomic molecules AlH with

$A = 44597\ \text{cm}^{-1}$	$\bar{\nu}' = 1326\ \text{cm}^{-1}$	$\bar{\nu} = 1625\ \text{cm}^{-1}$
$B' = 6.37\ \text{cm}^{-1}$	$\text{B} = 6.30\ \text{cm}^{-1}$	Assume $\Omega' = \Omega = 0$

Draw the bands for ($v' = 1$, $v = 0$) and ($v' = 2$, $v = 0$).

57. Look up and find data on two examples of alternation of intensities in the band spectra of homonuclear diatomic molecules (not including H_2 or O_2).

58. Plan an experiment of the Stern-Gerlach type to measure the magnetic moment of the H atom in its lowest quantum state. Consider all the essential elements: source, magnetic field shape, detector. Give quantitative specifications.

Collision theory, p. 145

59. Calculate the differential cross section for scattering from a potential having spherical symmetry and a dependence on r

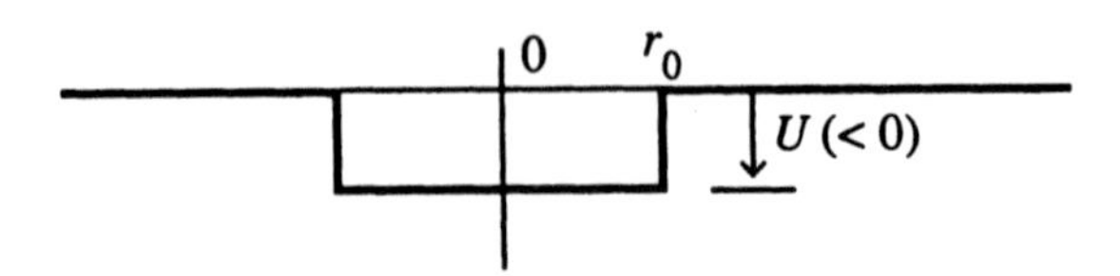

60. The scattering center is a (fixed) nucleus, $V(r) = -(Ze^2/r)$. Calculate the differential cross section for Coulomb scattering. To avoid troubles with convergence, use $V(r) = -(Ze^2/r)e^{-\lambda r}$ and let $\lambda \to 0$ after integration. The result is the Rutherford scattering formula.

Lightning Source UK Ltd.
Milton Keynes UK
UKOW02f0210171116
287805UK00003B/135/P

9 780226 243818